Genetic Engineering

Principles and Methods

Volume 1

GENETIC ENGINEERING
Principles and Methods

Advisory Board

A Continuation Order Plan is available for this series. A continuation order will bring delivery of each new volume immediately upon publication. Volumes are billed only upon actual shipment. For further information please contact the publisher.

Genetic Engineering

Principles and Methods

Volume 1

Edited by

Jane K. Setlow

Brookhaven National Laboratory
Upton, New York

and

Alexander Hollaender

Associated Universities, Inc.
Washington, D.C.

Plenum Press · New York and London

ISBN 0-306-40154-1

© 1979 Plenum Press, New York
A Division of Plenum Publishing Corporation
227 West 17th Street, New York, N.Y. 10011

Printed in the United States of America

Preface

This volume is the first of a series concerning a new technology which is revolutionizing the study of biology, perhaps as profoundly as the discovery of the gene. As pointed out in the introductory chapter, we look forward to the future impact of the technology, but cannot see where it might take us. The purpose of these volumes is to follow closely the explosion of new techniques and information that is occurring as a result of the newly-acquired ability to make particular kinds of precise cuts in DNA molecules. Thus we are particularly committed to rapid publication.

Jane K. Setlow

Alexander Hollaender

Contents

INTRODUCTION AND HISTORICAL BACKGROUND   1
    Maxine F. Singer

CLONING OF DOUBLE-STRANDED cDNA 15
    Argiris Efstratiadis and Lydia Villa-Komaroff

GENE ENRICHMENT 37
    M.H. Edgell, S. Weaver, Nancy Haigwood and
    C.A. Hutchison III

TRANSFORMATION OF MAMMALIAN CELLS 51
    M. Wigler, A. Pellicer, R. Axel and S. Silverstein

CONSTRUCTED MUTANTS OF SIMIAN VIRUS 40 73
    D. Shortle, J. Pipas, Sondra Lazarowitz, D. DiMaio
    and D. Nathans

STRUCTURE OF CLONED GENES FROM XENOPUS: A REVIEW 93
    R.H. Reeder

TRANSFORMATION OF YEAST 117
    Christine Ilgen, P.J. Farabaugh, A. Hinnen,
    Jean M. Walsh and G.R. Fink

THE USE OF SITE-DIRECTED MUTAGENESIS IN REVERSED GENETICS . . 133
    C. Weissmann, S. Nagata, T. Taniguchi, H. Weber and
    F. Meyer

AGROBACTERIUM TUMOR INDUCING PLASMIDS: POTENTIAL VECTORS
FOR THE GENETIC ENGINEERING OF PLANTS 151
    P.J.J. Hooykaas, R.A. Schilperoort and A. Rörsch

THE CHLOROPLAST, ITS GENOME AND POSSIBILITIES FOR
GENETICALLY MANIPULATING PLANTS 181
    L. Bogorad

MITOCHONDRIAL DNA OF HIGHER PLANTS AND GENETIC ENGINEERING .   205
    C.S. Levings III and D.R. Pring

HOST-VECTOR SYSTEMS FOR GENETIC ENGINEERING OF HIGHER
PLANT CELLS   223
    C.I. Kado

SOYBEAN UREASE-POTENTIAL GENETIC MANIPULATION OF
AGRONOMIC IMPORTANCE   241
    J.C. Polacco, R.B. Sparks Jr. and E.A. Havir

ACKNOWLEDGMENTS   261

INDEX   263

INTRODUCTION AND HISTORICAL BACKGROUND

Maxine F. Singer

Laboratory of Biochemistry, National Cancer Institute

Bethesda, Maryland 20014

Not even the wisest minds can reliably predict the future course of science. One particularly striking example of utter failure is a report on future technological developments prepared in 1937 by the best talent presumed to be available in the United States (1). The report concluded, for example, that any further changes in commercial aviation would be in the direction of improved safety and comfort, although the possibility was entertained that a limited number of airplanes capable of flying at 20,000 feet and at a speed of 240 miles per hour would be built. There is no mention whatever of jet engines nor, in other spheres, of plastics or nuclear energy. It seems that the fantastic imaginations of writers of science fiction are better at prediction than are the rational minds of experts.

Modern biologists cannot expect to be better seers than were their predecessors. This chapter introduces a book that describes recent experimental approaches to the manipulation of genes in the laboratory. How many readers of the book would have predicted its scope, contents, or even its existence, six years ago? And yet, paradoxically, today the contents of the book seem neither strange nor foreign. It is in fact rather simple to trace, in retrospect, the origins of the ideas and of the technical innovations. Even the results follow logically, albeit surprisingly, from all that came before. The origins are found among observations and concepts in genetics, enzymology, molecular biology, cell biology, botany, virology, and chemistry over about the past 35 years. And the framework of the present work is defined by two already venerable paradigms (2) -- the DNA revolution and the enzyme revolution.

While the revolution in biology that crystallized around the DNA structure is only now widely perceived by interested laymen-- partly as a result of the experiments described in this volume--

it is already a quarter of a century old. It is difficult to fix
a precise date for its start but the initial elements surely include the proof that DNA is the genetic substance (3,4), the determination of the structure of DNA (5), and the elucidation of
the genetic code (6-8). (The history of the DNA revolution has
recently been described in detail (9)). Most readers of this
book either internalized the revolution, bit by bit as it took
place, or else, if they are young enough, were educated in the
post-revolutionary period and were unaware of any cataclysm.

By itself, the DNA revolution was insufficient to permit
detailed and designed manipulation of genetic systems. Given
our customary time frame, it is difficult to recall that the
discovery of enzymes also constituted a revolution. The enzyme
revolution is as central to the successful manipulation of biological systems as is the DNA revolution. It began with the then
remarkable demonstration that complex macromolecular proteins
could be obtained from living organisms in the form of pure chem-
icals and yet still demonstrate those catalytic properties by
which they contributed to the "life" of the organism. (The history of the enzyme revolution has also been described recently
(10)). For the present purposes, the most fruitful development
within the framework of the enzyme revolution was the discovery
that other macromolecules, including polynucleotides, are substrates for specific enzymes. Chemistry itself has not yet
provided techniques for the precise manipulation of biological
macromolecules. Without the enzymes, the subject matter of this
book would not now be attainable.

Why then, if the concepts were familiar and the paradigms
of long standing, was prediction of the present results so unlikely? To a large extent, the lack of predictability stems
from the diverse provenance of the antecedents of the present
experiments. Furthermore, the timing of the necessary prior
discoveries was not colinear with the logical development of the
new methods. The two revolutions provided remarkably fruitful
frameworks for the interpretation of biological observations,
especially as they began to converge. The ability to move beyond
interpreting observable phenomena and actually to manipulate
biological systems in the fundamental manner described in this
book, could not be foreseen.

With very few exceptions, the readers of this book were
educated in the post-revolutionary period as far as the enzyme
revolution is concerned. They will not usually concern themselves with the remarkable nature of the enzymes used to manipulate and construct precise DNA molecules. Most often those
enzymes are perceived only as tools and are confronted in
practice as a rather disorganized array of odd shaped tiny tubes
inside a freezer. The sight inside the freezer carries no re-
minder that the fruits of a revolution are at hand. Nor will
the investigator be reminded by the means he uses to acquire
the enzymes. Some of them are now as easy to come by all over

the world as a bottle of Coca-Cola—only more expensive. Others
may be obtained by persuasive begging and borrowing, and, in a
tight squeeze, even stealing. But when all else fails, and it is
necessary actually to prepare one of the enzymes, the impact of
the revolution may be sensed. Delight and amazement inevitably
accompany the emergence of a clean, exquisitely specific reagent
from the equally inevitably messy beginnings. Dismay coupled
with awe of successful predecessors accompany the frequent
frustrating failures.

    Current emphasis on the utility of reagents to manipulate
DNA has also obscured the inherent and singular importance of
many of the enzymes. They represent the common ground of the
DNA revolution and the enzyme revolution, and in many instances
their discoveries added facts of first magnitude importance to
our knowledge of biology. The role of many of these enzymes in
the growth and reproduction of cells is a central issue for
modern biology. Study of the mechanism behind the complex yet
highly specific reactions they catalyze is still a major concern
of modern enzymologists. It is interesting and instructive to
review the history of the discovery of these enzymes and to re-
mind ourselves of their significance.

THE ALKALINE PHOSPHATASE OF Escherichia coli

    The alkaline phosphatase of E. coli seems a very mundane
reagent, devoid of any scientific glamour at all. Phosphatases
catalyze a rather simple reaction, the hydrolysis of phosphomono-
esters. The enzyme from E. coli is not even very fastidious—
any phosphomonoester will do and even polyphosphates can be
hydrolyzed. Besides, E. coli alkaline phosphatase of satisfactory
purity has been available commercially for many years and it is
cheap. And yet its discovery by Horiuchi, Horiuchi and Mizuno
(11,12), and independently by Torriani (13), was of great interest in the late 1950s because it was made in the context of the
early efforts to understand regulation of cellular processes.
Earlier workers had observed that resting bacteria tend to degrade their own proteins and nucleic acids. Attempting to define
this phenomenon more carefully, Torriani, as well as Horiuchi
and coworkers, starved E. coli cells for single components.
When the cells were starved for phosphate they did indeed degrade
RNA but continued to make DNA and protein and to grow and multiply. A survey of the levels of a variety of enzymes under such
conditions revealed a marked increase in phosphomonoesterase
activity. The increased activity resulted from new synthesis
of a single enzyme—the alkaline phosphatase. Under conditions
of phosphate starvation this enzyme accounts for as much as 6%
of the cellular protein (14). Even more dramatically, the increase in activity stopped promptly upon addition of phosphate
to the cells. Thus synthesis of alkaline phosphatase is regulated

by a negative feedback system in which inorganic phosphate is the regulator. Quite separately, inorganic phosphate also interacts directly with alkaline phosphatase and inhibits its activity (14). The cell can thus respond to inorganic phosphate both in an immediate mode (inhibition of enzymatic activity) and in a long term mode (repression of enzyme synthesis).

Accompanying Torriani's paper is a report by Garen and Levinthal (14) on the purification of the phosphatase from derepressed cells. Having learned about the enzyme from their colleague Torriani, they recognized its potential for correlating changes in enzyme structure brought about by mutation, with altered susceptibility to regulation. The system still seems an attractive one, though in fact it has never been fully exploited.

Garen and Levinthal showed that alkaline phosphatase was a general, nonspecific phosphatase (14). Khorana and Viszolyi first reported that the enzyme could dephosphorylate polynucleotides (15). Fortunately, for the present purposes, Richardson (16) showed that the difficulty the enzyme has in removing phosphate from monoester ends that are either at nicks in duplex DNA, or otherwise covered by an overhanging complementary strand (first observed at the 5'-terminus of transfer RNA (17)) can be overcome by carrying out the dephosphorylation at elevated temperature.

S1 NUCLEASE

The rapid breakdown of DNA and RNA in all sorts of crude extracts is a recurrent and often vexing side effect in experiments designed for other purposes. In not a few instances during the last 25 years, purification of a nuclease became the only constructive outcome of otherwise frustrating results. Determined searches for nucleases were of course always successful. At times it seemed as though too many people and too many journal pages were devoted to the description of yet another nuclease. Nucleases were described from sources such as shark liver and mung bean and everything in between. But finally, what seemed like redundant, even trivial, investigations proved remarkably fruitful. Many of the enzymes exhibited uniquely different approaches to the degradation of polynucleotides. In recent years, these differences have been exploited to produce elegant and specific analytical procedures.

Takadiastase, a dried powder made from _Aspergillus_ _oryzae_, was one of the odd sources searched for nuclease. This material is prepared in large quantity in Japan where it is commonly used to alleviate human digestive ills. It had already proven to be a convenient source of amylase, when, in 1966, Ando detected in it a nuclease now known as S1 (18). When the enzyme was purified, first by Sutton (19) and later by Vogt (20), it turned out to have a marked and very useful preference for single-stranded polynucleotides. A similar enzyme had been purified previously from

Neurospora crassa (21) but the ease with which S1 can be obtained
makes it the enzyme of choice.

LAMBDA EXONUCLEASE

The availability of simple methods for the detecting of
nucleases makes these enzymes ideal potential indicators of changes
in the physiological state of cells. By the early 1960s, it was
known that infection of E. coli with any of a number of virulent
bacteriophages results in the formation of new enzymatic activities, including nucleases, and the mapping of the genes for these
enzymes on the bacteriophage chromosome had begun. At the same
time, some believed that lysogenic induction of cells lysogenized
by those same bacteriophage did not result in new enzymatic activities. Korn and Weissbach (22,23) were interested in testing
this hypothesis and chose to measure nuclease activity after induction of E. coli cells lysogenized with bacteriophage lambda.
A large increase in DNase activity was readily detected and
proved to be both encoded by the lambda genome and identical to
the activity found in virulent infections. When the enzyme was
purified to homogeneity and crystallized by Little, Lehman, and
Kaiser (24), its mode of action was confirmed and it was dubbed
"lambda exonuclease" (25). The products are 5'-nucleoside monophosphates (25) and, as the very first experiments suggested (22,
23), the enzyme prefers double-stranded DNA as a substrate (25,26).
The lambda exonuclease turned out to be particularly interesting
both mechanistically and as an analytical reagent when Little (25)
determined that its exonucleolytic cleavage of DNA started at the
5'-terminus of the chain. Until that time, the only nuclease
known to proceed in the 5' to 3' direction was the exonuclease of
spleen, which cleaves both DNA and RNA and produces 3'-nucleoside
monophosphates (27,28). The spleen enzyme was notoriously difficult to separate from relevant contaminating activities and
therefore of limited analytic potential. Subsequently, the 5'
to 3' exonuclease activity of DNA polymerase I was discovered (29).
The lambda exonuclease is not yet available commercially.
When the enzyme is needed, the molecular biologist has an opportunity to test his mettle as enzymologist. The test is not however very rigorous. Thanks to the work of Radding and his associates (30,31), the enzyme can be prepared from induced lysogens
that superproduce early lambda proteins, including lambda-exonu-
clease, to such an extent that only a 90-fold purification suffices
to yield pure enzyme (24).

RESTRICTION ENDONUCLEASES

The power of congruence of the DNA revolution and the enzyme
revolution is beautifully illustrated by the history of the

discovery of restriction endonucleases. Almost 30 years ago, the
early phage workers recognized that the ability of a bacteriophage
to reproduce in a particular cell type depends on the cell type in
which the phage was previously grown (32). The phenomenon was referred to as host controlled variation, or host induced restriction,
and appeared to be unrelated to the genetic makeup of the bacteriophage itself. To explain these odd findings, Bertani and Weigle
(33) speculated in 1953 that some bacteriophage component (unspec-
ified, but required for bacteriophage multiplication) was under
the control of the host. Later, Lederberg (34) learned that a
lysogenic bacteriophage could similarly restrict multiplication
of a heterologous bacteriophage. Lederberg obtained the first
experimental clue to the mechanism of restriction when he noted
that radioactivity from the labeled DNA of an unsuccessful heterologous bacteriophage appeared in the medium. Some five years
later, Arber and Dussoix (35-37) confirmed the relation between
restriction and DNA degradation in the course of extensive
studies on restriction. Utilizing a series of mutants, these
investigators elucidated the intricate and related mechanisms
behind restriction, which acts on an invading bacteriophage, and
its obverse modification, which permits recognition of self. In
1965, reviewing the earlier experimental work from his laboratory,
Arber (37) argued that the observed gross breakdown of the
restricted DNA was probably the result of a highly specific initial cleavage followed by subsequent, nonspecific degradation.
Since modification appeared to depend on methionine, Arber
speculated further that the modification that inhibits restriction
involves alkylation of the DNA. Though direct evidence was lacking, Arber saw it was most likely that the specificity of both
processes resided in the base sequence of the DNA. He wrote (37):

> "If this last idea should be correct, one may further
> speculate that a restriction enzyme might provide a
> tool for the sequence specific cleavage of DNA.
> Application of enzymes of different specificity
> should then be useful in attempts to determine base
> sequences of DNA molecules."

The idea was of course correct. By then, restriction-modifi-
cation systems were known to be programmed by bacterial chromosomes,
by bacteriophage chromosomes, and by plasmids. So Arber may well
have realized how large a catalog of enzymes of different specificities would ultimately be available (38). Nevertheless, five
years went by before Arber's predictions came true. During that
time, experimental work concentrated on the group of restriction
endonucleases now termed class I. These enzymes are fascinating
and still not completely understood, but they are not geared to
fine surgery of DNA as they do not cleave at specific sites.
Then, in 1970, Smith, with Wilcox (39) and with Kelly (40),
described the very specific cleavage of DNA at a given nucleotide

sequence by what are now called class II enzymes. Since then we
have had an avalanche of enzymes from a variety of esoteric bacteria, each cleaving its own favored base sequences. Molecular
biologists have been reminded that E. coli is not, so to speak,
the only fish in the sea. Any remaining doubts about the accuracy
of Arber's predictions were removed by Danna and Nathan's (41)
construction of a physical map of the genome of simian virus 40,
using the endo R·HindII and endo R·HindIII discovered by Smith
and his colleagues.

The restriction endonucleases yielded other spectacular,
though unsuspected gifts. The discovery by Mertz and Davis (42)
and Hedgpeth, Goodman, and Boyer (43) that the cleavage products
produced by endo R·EcoRI have single-stranded, complementary,
overhanging ends, and subsequent work showing that other enzymes
did too, greatly improved the prospects for joining DNA fragments.

DNA LIGASE

The discovery of DNA ligase was one of the greatest "happen-
ings" of 1967. It was another event important to the growing
congruence of the DNA revolution and the enzyme revolution; and
the atmosphere was enlivened because the discovery was made essentially simultaneously by several independent groups working in the
laboratories of Gellert, Richardson, Lehman, Hurwitz, and Kornberg
(44-48).

At the time, the most attractive proposals concerning the
mechanism of genetic recombination postulated both breaking and
rejoining DNA chains. Mechanisms for breakage abounded in the
many demonstrable nucleases, but rejoining remained a speculative
process until ligase was discovered. The importance of the
enzyme was emphasized by its presence in E. coli itself, and by
the synthesis of a new ligase upon infection of cells with bacteriophage T4, although the role of DNA ligase in replication
itself was not yet suspected. The E. coli and T4 enzymes differ
significantly in two ways. First, the E. coli ligase requires
diphosphopyridine nucleotide as a cofactor while the T4 ligase
requires adenosine triphosphate even though each uses its cofactor to form analogous, ligase-adenylate and DNA-adenylate intermediates (49). Second, the T4 ligase does not require overlapping
complementarity on the single-stranded ends of the two chains to
be joined as does the E. coli enzyme (50). Two DNA molecules
with fully complementary chains can be readily joined by T4
enzyme in what is referred to as "blunt end" ligation; the only
requirement is that there be a phosphomonoester end group at the
5'-terminus and a hydroxyl group at the 3'-terminal.

DNA ligase is central to the recombination of DNA in vitro.
It is satisfying to realize that one of the earliest predicted
results of recombinant DNA research, the construction of bacteria

that would be efficient sources of important proteins, was first
realized with DNA ligase itself (51,52).

DNA POLYMERASE I

One of the most dramatic results of the juxtaposition of the
DNA and enzyme revolutions was the discovery of enzymes and enzyme
systems that would, outside of whole cells, copy polynucleotide
templates and synthesize large, specific macromolecules. Cell-
free synthesis of DNA, of RNA, and of proteins were described in
rapid succession between 1958 and 1962. Discovery of the DNA
polymerase I of E. coli came first, and its discovery by Kornberg
and his colleagues was not accidental (53,54). Educated and
accurate guesses were based on the proposed structure of DNA and
its inherent implications about its own replication (5) as well
as on the insight into transfer reactions involving polyphosphate
esters of nucleotides that had been obtained from the earlier
elucidation of the biosynthesis of nucleotides and coenzymes.
Now, 20 years later, we realize that the synthesis of DNA is more
complex and more diverse than might have been imagined in 1958.
Still, the description of DNA polymerase I (54) and most importantly,
the ability of the enzyme to copy a template faithfully (53,55),
opened a new era.

Many other DNA polymerases are now known, but polymerase I
remains among the most interesting, perhaps because it is the best
understood. From an enzymologist's viewpoint, the interrelation
between the three quite different reactions catalyzed by the two-
headed protein remain intriguing. One head catalyzes phosphoryl
transfer and thus synthesis, as well as hydrolysis in the 3' to 5'
direction (53). The other head catalyzes hydroysis in the 5' to 3'
direction (29,53), a reaction analogous to that catalyzed by lambda
exonuclease. The two heads together, under the proper conditions,
can start a nick in a DNA duplex, degrade the chain in the 5' to
3' direction and simultaneously rebuild the degraded chain by addition to the 3'-hydroxyl that is on the other side of the nick
(53,56). The nick thus progresses down the chain--a process commonly and perhaps unfortunately referred to as nick translation,
rather than nick progression. Notwithstanding the oddity of the
name, the process itself is extraordinarily useful--it allows the
preparation of ^{32}P-labeled DNA fragments with specific radioactiv-
ities of the order of 2×10^8 counts per minute per microgram (57).

The E. coli gene for DNA polymerase I has also been amplified
in E. coli by means of recombinant DNA techniques (58).

TERMINAL NUCLEOTIDYL TRANSFERASE

The discovery of DNA polymerase I stimulated a search for
similar enzymes in eukaryote sources. The result was a bonanza

and it was quickly recognized that more than one DNA polymerase
coexist in cells. The polymerizing activity that was readily
detected by Bollum in extracts of calf thymus was initially believed to be a DNA polymerase (59). The early work on this
activity concentrated on the fact that the enzyme catalyzed the
addition of deoxymononucleotide residues to a preformed poly-
deoxyribonucleotide primer and preferred single-stranded primers
for the reaction (60). Ultimately, of course, it turned out that
all the DNA polymerases require a primer. But, in contrast to
the true polymerases, the major activity in calf thymus extracts
does not require both a template and a primer and was eventually
renamed terminal nucleotidyl transferase to reflect its mechanism
of action (61–63). Calf thymus extracts contain other enzymes
that can polymerize deoxynucleotide triphosphates by copying a
template (63).

The preference for single-stranded primers is the basis for
the need to digest DNA fragments briefly with lambda exonuclease
prior to the addition of residues with terminal nucleotidyl
transferase (64,65).

REVERSE TRANSCRIPTASE

Reverse transcriptase is another of the enzymatic reagents
that should be appreciated as a major landmark in the process of
joining the two revolutions. Data published by Temin as early as
1964 (66) suggested that such an enzyme might exist and in 1970
the predicted activity was demonstrated in preparations from RNA
tumor viruses by Temin and his colleagues (67) and by Baltimore
(68). Some observers saw the discovery as somehow detracting
from the central position of DNA as the repository for biological
information. But if the finding is viewed another way, the
essential role of the enzyme in the replication of certain RNA
viruses, if anything, confirms the central position of DNA in
biology. Besides, now that it is recognized that the genomes
of RNA viruses quite commonly reside within the genomes of cells
in the form of DNA, the question of the relative importance of
DNA and RNA is reduced to the old chicken and egg problem. In
any case, the addition of reverse transcriptase to the catalog
of known enzymes made a two-directional flow of information--DNA
to RNA and RNA to DNA--feasible both in cells and in test tubes.

To carry out the RNA to DNA conversion in test tubes in a
preparatively useful manner requires the availability of sufficiently purified enzyme in adequate quantities. With some
variation, the large scale procedures introduced by Spiegelman
in 1972 (69) still provide the bulk of the needs of investigators.

CONCLUSION

This is an introduction to a book and not meant to be an exhaustive review. Many enzymes that are being used and will be used for the manipulation of genes and chromosomes have not been mentioned. Still the histories of these few enzymes suffice to illustrate that the future impact of individual discoveries will rarely, if ever, be completely predictable. Each new discovery is likely to have pleiotropic effects and some of these effects will be apparent only later on, in the context of subsequent discoveries and theories or even new paradigms. The people who discovered the enzymes used for the manipulation of DNA did not spend time speculating on what they knew they could not yet know. Similarly, while the investigators reporting work in this book are delighted with what they have learned, neither they nor anyone else can say with certainty what may be learned in the future. The present authors will, of course, speculate; they would be neither human nor scientists if they did not. Predictions and speculations form the basis for new experiments, and some predictions, like those of Arber, may turn out to be correct.

For most scientists the frustration of not knowing the future is neither debilitating nor cause for alarm. Pessimism is rejected because it is both uninteresting and unproductive. Indeed, one might define science itself as the optimist's response to the conundrum expressed so well by the novelist and poet Robert Penn Warren (70).

> "The end of man is knowledge, but there is one thing he can't know. He can't know whether knowledge will save him or kill him. He will be killed, all right, but he can't know whether he is killed because of the knowledge which he has got or because of the knowledge which he hasn't got and which if he had it, would save him."

Acknowledgments: I thank Harold P. Green for bringing the 1937 report on Technological Trends (1) to my attention. It is a great pleasure to acknowledge here my personal debt to three people who taught me to appreciate the magnitude and wonder of the enzyme revolution: Joseph S. Fruton, Ephraim Racker, and Leon A. Heppel.

REFERENCES

1   Technological Trends and National Policy, Report of the Sub-
    committee on Technology to the U.S. National Resources Com-
    mittee (1937) U.S. Government Printing Office, Washington, D.C.
2   Kuhn, T.S. (1962) The Structure of Scientific Revolutions,
    The University of Chicago Press, Chicago, IL.

3   Avery, O.T., MacLeod, C.M. and McCarty, M. (1944) J. Exp. Med.
     79, 137-158.
4   Hershey, A.D. and Chase, M. (1952) J. Gen. Physiol. 36, 39-56.
5   Watson, J.D. and Crick, F.H.C. (1953) Nature 171, 737-738.
6   Nirenberg, M.W. and Matthaci, J.H. (1961) Proc. Nat. Acad. Sci.
     U.S.A. 47, 1588-1602.
7   Leder, P. and Nirenberg, M. (1964) Proc. Nat. Acad. Sci. U.S.A.
     52, 420-427.
8   Nishimura, S., Jones, D.S. and Khorana, H.G. (1965) J. Mol.
     Biol. 13, 302-324.
9   Olby, R. (1974) The Path to the Double Helix, University of
     Washington Press, Seattle, WA.
10   Fruton, J.S. (1972) Molecules and Life, Wiley-Interscience,
     New York, NY.
11   Horiuchi, T., Horiuchi, S. and Mizuno, D. (1959) Biochim.
     Biophys. Acta 31, 570-572.
12   Horiuchi, T., Horiuchi, S. and Mizuno, D. (1959) Nature 183,
     1529-1530.
13   Torriani, A. (1960) Biochim. Biophys. Acta 38, 460-469.
14   Garen, A. and Levinthal, C. (1960) Biochim. Biophys. Acta 38,
     470-483.
15   Khorana, H.G. and Vizsolyi, J.P. (1961) J. Amer. Chem. Soc.
     83, 675-685.
16   Richardson, C.C. (1966) J. Mol. Biol. 15, 49-61.
17   Harkness, D.R. and Hilmoe, R.J. (1962) Biochem. Biophys. Res.
     Commun. 9, 393-397.
18   Ando, T. (1966) Biochim. Biophys. Acta 114, 158-168.
19   Sutton, W.D. (1971) Biochim. Biophys. Acta 240, 522-531.
20   Vogt, V.M. (1973) Eur. J. Biochem. 33, 192-200.
21   Linn, S. and Lehman, I.R. (1965) J. Biol. Chem. 240, 1287-1293.
22   Weissbach, A. and Korn, D. (1962) J. Biol. Chem. 237,
     3312-3314.
23   Korn, D. and Weissbach, A. (1963) J. Biol. Chem. 238,
     3390-3394.
24   Little, J.W., Lehman, I.R. and Kaiser A.D. (1967) J. Biol.
     Chem. 242, 672-678.
25   Little, J.W. (1967) J. Biol. Chem. 242, 679-686.
26   Radding C.M. (1966) J. Mol. Biol. 18, 235-250.
27   Hilmoe, R.J. (1960) J. Biol. Chem. 235, 2117-2121.
28   Razzell, W.E. and Khorana, H.G. (1961) J. Biol. Chem. 236,
     1144-1149.
29   Klett, R.P., Cerami, A and Reich, E. (1968) Proc. Nat. Acad.
     Sci. U.S.A. 60, 943-950.
30   Radding, C.M. (1964) Proc. Nat. Acad. Sci. U.S.A. 52, 965-973.
31   Radding, C.M. and Shreffler, D.C. (1966) J. Mol. Biol. 18,
     251-261.
32   Luria, S.E. and Human, M.L. (1952) J. Bacteriol. 64, 557-569.
33   Bertani, G. and Weigle, J.J. (1953) J. Bacteriol. 65, 113-121.
34   Lederberg, S. (1957) Virology 3, 496-513.

35 Arber, W. and Dussoix, D. (1962) J. Mol. Biol. 5, 18-36.
36 Dussoix, D. and Arber, W. (1962) J. Mol. Biol. 5, 37-49.
37 Arber, W. (1965) Ann. Rev. Microbiol. 19, 365-378.
38 Roberts, R.J. (1976) CRC Critical Rev. Biochem. 4, 123-164.
39 Smith, H.O. and Wilcox, K.W. (1970) J. Mol. Biol. 51, 371-391.
40 Kelly, T.J. Jr. and Smith, H.O. (1970) J. Mol. Biol. 51, 393-409.
41 Danna, K.J. and Nathans, D. (1971) Proc. Nat. Aca. Sci. U.S.A. 68, 2913-2917.
42 Mertz, J.E. and Davis, R.W. (1972) Proc. Nat. Acad. Sci. U.S.A. 69, 3370-3374.
43 Hedgpeth, J., Goodman, H.M. and Boyer, H.W. (1972) Proc. Nat. Acad. Sci. U.S.A. 69, 3448-3452.
44 Gellert, M. (1967) Proc. Nat. Acad. Sci. U.S.A. 57, 148-155.
45 Weiss, B. and Richardson, C.C. (1967) Proc. Nat. Acad. Sci. U.S.A. 57, 1021-1028.
46 Olivera, B.M. and Lehman, I.R. (1967) Proc. Nat. Acad. Sci. U.S.A. 57, 1426-1433.
47 Gefter, M.L. Becker, A. and Hurwitz, J. (1967) Proc. Nat. Acad. Sci. U.S.A. 58, 240-247.
48 Cozzarelli, N.R., Melechen, N.E., Jovin, T.M. and Kornberg, A. (1967) Biochem. Biophys. Res. Commun. 28, 578-586.
49 Lehman, I.R. (1974) Science 186, 790-797.
50 Sgaramella, V., Van de Sande, J.H. and Khorana, H.G. (1970) Proc. Nat. Acad. Sci. U.S.A. 67, 1468-1475.
51 Panasenko, S.M., Cameron, J., Davis, R.W. and Lehman, I.R. (1977) Science 196, 188-189.
52 Panasenko, S.M., Alazard, R.J. and Lehman, I.R. (1978) J. Biol. Chem. 253, 4590-4592.
53 Kornberg, A. (1974) DNA Synthesis, W.H. Freeman and Company, San Francisco, CA.
54 Lehman, I.R., Bessman, M.J., Simms, E.S. and Kornberg, A. (1958) J. Biol. Chem. 233, 163-170.
55 Josse, J., Kaiser, A.D. and Kornberg, A. (1961) J. Biol. Chem. 236, 864-875.
56 Kelly, R.B., Cozzarelli, N.R., Deutscher, M.P., Lehman, I.R. and Kornberg, A. (1970) J. Biol. Chem. 245, 39-45.
57 Rigby, P.W.J., Dieckmann, M., Rhodes, C. and Berg, P. (1977) J. Mol. Biol. 113, 237-251.
58 Kelley, W.S., Chalmers, K. and Murray, N.E. (1977) Proc. Nat. Acad. Sci. U.S.A. 74, 5632-5636.
59 Bollum, F.J. (1960) J. Biol. Chem. 235, PC18-PC20.
60 Bollum, F.J. (1962) J. Biol. Chem. 237, 1945-1949.
61 Krakow, J.S., Coutsogeorgopocilos, C. and Canellakis, E.S. (1962) Biochim. Biophys. Acta 55, 639-650.
62 Keir, H.M. and Smith, M.J. (1963) Biochim. Biophys. Acta 68, 589-598.
63 Yoneda, M. and Bollum, F.J. (1965) J. Biol. Chem. 240, 3385-3391.
64 Jackson, D.A., Symons, R.H. and Berg, P. (1972) Proc. Nat. Acad. Sci. U.S.A. 69, 2904-2909.

65   Lobban, P.E. and Kaiser, A.D. (1973) J. Mol. Biol. 78, 453–471.
66   Temin, H.M. (1964) Virology 23, 486–494.
67   Temin, H.M. and Mizutani, S. (1970) Nature 226, 1211–1213.
68   Baltimore, D. (1970) Nature 226, 1209–1211.
69   Kacian D.L. and Spiegelman S. (1972) in Methods in Enzymology,
      Vol. 29, pp. 150–173, Academic Press, New York, NY.
70   Robert Penn Warren (1946) All the King's Men, Harcourt, Brace
      and Company, New York, NY.

CLONING OF DOUBLE-STRANDED cDNA

Argiris Efstratiadis

Department of Biological Chemistry
Harvard Medical School
Boston, Massachusetts   02115

and

Lydia Villa-Komaroff

Department of Microbiology
University of Massachusetts Medical Center
Worcester, Massachusetts   01605

INTRODUCTION

The structural and functional characteristics of specialized eukaryotic cells depend on the production of specific proteins by regulated differential expression of only a portion of the genetic information. It seems that a fundamental part of this process in eukaryotes (as in prokaryotes) is regulation at the level of transcription, although control mechanisms could operate at any step in the flow of genetic information.

The understanding of the molecular basis of transcriptional control presupposes extensive information about the structure and organization of specific genes and their associated sequences in the genome. Such information can be gained by using as model systems highly differentiated cells, specialized in the production of a few, well characterized proteins. If the mRNAs corresponding to these proteins can be purified, they can serve as probes to identify the respective structural genes. Therefore, such systems offer the opportunity of a direct biochemical study of the chromosomal arrangement of specific genes. This approach, despite limitations due to the general unavailability of mutants, is particularly important because very few, well defined genetic systems are amenable to biochemical analysis.

The complexity of the eukaryotic genome precludes, in general, the direct purification of single-copy structural genes by conventional methods. However, purification and further study can be accomplished by DNA cloning, using two methods that complement each other: the construction of libraries of eukaryotic DNA and the cloning of double-stranded cDNA (ds-cDNA).

Libraries are formed by cloning of random fragments of chromosomal DNA (generated by shearing or by specifically designed partial restriction endonuclease digestion). The library is complete if the number of derived clones is large enough for complete sequence representation (1). Complete libraries have been constructed from total chromosomal DNA of several organisms with a wide spectrum of genome sizes (1-3). Partial libraries have been constructed by enrichment of mammalian DNA for interesting genes prior to cloning (4,5). In principle, any gene of interest can be isolated from the library by the use of a specific hybridization probe.

Double-stranded cDNA (ds-cDNA) technology was initially developed to fulfill the need for such probes. Its use, however, is not limited to this application. In practice, the generation of pure probes is often a very difficult biochemical task, particularly when many different mRNA species are present in a system, some of them in extremely small amounts. One way (and in certain cases the only way) of isolating individual sequences in useful amounts is to convert the entire mRNA population of a system into ds-cDNA and produce homogeneous probes through cloning. This is primarily the reason that, in the study of a system, ds-cDNA cloning usually precedes the construction of a library. In conjunction with direct RNA sequencing, cloned ds-cDNA can be used in sequencing studies for the determination of the primary structure of an mRNA. Since the coding region of eukaryotic structural genes is often interrupted by introns (6-8), knowledge of an mRNA sequence is essential for comparison with that of the corresponding gene after its isolation from a library. Moreover, mRNA sequence information is important for evolutionary studies (e.g., ref. 9) and studies concerning secondary structure or the function of the noncoding regions (e.g., ref. 10). Double-stranded cDNA can also be employed in studies concerning the expression of eukaryotic sequences in bacteria. In this respect, it has an obvious advantage over cloned chromosomal DNA because introns, if present in eukaryotic structural genes, will be transcribed and bacteria, presumably lacking RNA processing systems, will be unable to form mature mRNA.

METHODOLOGY

Double-stranded cDNA technology was developed using rabbit globin mRNA as the starting material because at the time it was the best characterized eukaryotic mRNA available in substantial amounts. The molecular weights of the α- and β-globin mRNAs had been determined (11), portions of them had been sequenced (12,13), and the amino acid sequences of the corresponding globin chains were known (14).

All variations of the procedures developed for ds-cDNA cloning start by synthesis of a DNA copy of the mRNA (cDNA) using reverse transcriptase.

Strictly speaking, one of the methods does not involve ds-cDNA synthesis. Instead a poly(dA)-tailed mRNA:cDNA heteroduplex was cloned after insertion in the EcoRI site of plasmid ColEl (15). Evaluation of the methodology in this report is difficult but if the acceptance of an RNA:DNA heteroduplex by a plasmid is a general phenomenon, the method would be very appealing because of its simplicity. Unfortunately, the efficiency of this procedure is low (16).

In another, rather complicated approach (17), a poly(dT) tail added to a linearized plasmid was used to prime the copying of mRNA by reverse transcriptase. Following addition of poly(dT) tails to the synthesized cDNA, the duplexes carrying globin sequence were dissociated and annealed to poly(dA)-tailed plasmids.

In another method (18), the cDNA transcript was elongated with a homopolymeric tract using terminal transferase and the second strand was synthesized in the presence of a complementary oligonucleotide primer with E. coli DNA polymerase I (Pol-I).

In the self-priming method (19-22), described in detail below and now the most commonly used, the cDNA serves both as template and primer for the synthesis of the second strand which is covalently linked to the first. The loop of the hairpin molecule formed can be specifically cleaved with single-strand-specific S1 nuclease.

Synthesis of the First Strand

RNA tumor viruses contain an RNA-directed DNA polymerase (reverse transcriptase, RT) (23-25). Although other polymerases, including Pol-I, can transcribe a primed RNA template under certain in vitro conditions (26-29), the readily available RT from avian myeloblastosis virus (AMV-RT) is widely used for this purpose because of its greater transcriptional efficiency. In 1972, three laboratories independently described the synthesis of a single-stranded DNA copy (cDNA) of globin mRNA using AMV-RT (30-32). The cDNA transcripts of various mRNAs were soon successfully employed for sequencing studies and as hybridization probes, substituting for mRNA because of the higher specific activities obtainable and the greater stability of DNA. However, when the cDNA represents transcripts of a heterogeneous mRNA population, the interpretation of mRNA:cDNA hybridization kinetics for defining abundance classes is based on the assumption (33) that transcriptional efficiency is equal for all templates. This may not always be true. For example, of the oviduct mRNAs (encoding ovalbumin, conalbumin, ovomucoid and

lysozyme), the conalbumin template is relatively inefficient (34).
Fibroin mRNA is another example of a particularly inefficient
template (35).

Since the first step in ds-cDNA synthesis is reverse transcription, adjustment in each case of the reaction conditions for synthesis of full length transcripts at high yield seems necessary.
The experimental conditions for cDNA synthesis have been reviewed
in detail (36,37). Here, we will focus on certain important points.
Optimization studies are not easy because of the many parameters
involved in the reaction. In some cases, for example, availability
of the mRNA in small amounts will be a serious limiting factor. In
other instances, even exhaustive optimizations might lead to con-
tradictory results (compare e.g., refs. 34 and 38). Therefore,
keeping in mind that the goal is not the ideal reverse transcription
but the final production of ds-cDNA as close as possible to full
length in clonable amounts (a reasonable minimum is of the order of
0.05-0.1 µg), we suggest the following strategy which compromises
between the variables, including cost and time.

First, control experiments (often overlooked) are necessary to
examine the purity of each enzyme preparation. Though the various
purification schemes (39-42) seem to yield RT devoid of DNase con-
taminants, the presence of RNase (see e.g., refs. 34 and 43), probably endoribonuclease (44), in variable amounts between preparations
even from the same source, is probable. This will cause serious
problems, especially with mRNA templates longer than 1000 nucleotides (NT). A sensitive assay for RNase contamination is exposure
of homogeneously ^{32}P-labeled rRNA to the enzyme under conditions of
reverse transcription followed by gel electrophoresis and autoradiography. As a test molecule, tRNA cannot substitute for rRNA
because it is relatively insensitive to RNase action. If RNase
activity is detected the enzyme should be further purified (40) or
RNase inhibitors (45) can be used if it is proved that they offer
protection without impairing reverse transcription. It is con-
ceivable that certain of the conditions used to increase the proportion of full length transcripts simply inhibit nuclease contam-
inants (see below).

Two factors that cause inhibition of RT should always be considered: a) Large amounts of a white residue which sometimes
appears upon drying of ^{32}P-deoxynucleotide triphosphates (dNTPs)
(possibly contaminating triethyl ammonium bicarbonate) inhibit the
enzyme. This contaminant can be removed by repeated lyophilizations.
b) If the mRNA is purified from a polyacrylamide gel, it should be
passed over an oligo(dT)-cellulose column before reverse transcription; RT is strongly inhibited by gel impurities (probably linear
polyacrylamide) which are invariably eluted together with the nucleic
acid, are ethanol precipitable, and cannot be removed by gel filtration or centrifugation.

The purity of the mRNA template to be employed for ds-cDNA
synthesis is not of primary importance. Since resolution of sequences will result from cloning, it is often desirable to reduce

to a minimum the number of purification steps prior to reverse transcription to avoid losses. Even mRNA bound to oligo(dT)-cellulose only once is generally an adequate template, despite the fact that it is contaminated with rRNA and tRNA (46) which are transcribed, but with very low efficiency (30). A single selection on oligo(dT)-cellulose is probably the scheme-of-choice for the reverse transcription of a heterogeneous mixture of mRNAs with a wide range of concentrations and lengths. Some length selection can be applied at a later stage (see below). In another section, we will discuss various solutions of a serious problem inherent in this approach: the screening of clones. The same scheme is also advisable when the mRNA of interest is a minor species in an mRNA population. In such cases, optimization studies are almost impossible, though length assays for the interesting sequence might be feasible by restriction endonuclease analysis (47,48) using enzymes known to specifically cleave single-stranded DNA (e.g., HaeIII). In general, however, if the mRNA is shorter than 1000 nucleotides, conditions established as optimal for the reverse transcription of globin mRNA can be applied with success (e.g., see refs. 43 and 44).

The composition of the enzyme storage buffer should always be considered, since it affects the composition of the transcription reaction mixture. Most importantly, precautions should be taken so that the final pH of the reaction is 8.3 (optimum), because the pH of the enzyme storage buffer is usually 7 to 7.5 (at pH 7.2, the incorporation is 50% of that at pH 8.3) (41).

Attention should be paid to the concentrations of divalent and monovalent cations, because the ionic conditions substantially affect the transcriptional efficiency of various templates (39,49). Except for their template-dependent activity, divalent cations are an absolute requirement for RT action. Mg^{2+} is almost invariably used for mRNA transcription. The enzymatic activity is lost at Mg^{2+} concentrations lower than 4 to 6 mM (38). However, optimization studies are not absolutely necessary because at a concentration of 10 mM, which was shown to be optimal for globin (36) and AMV (50) mRNAs, RT transcribes other mRNAs efficiently as well (34,44). On the other hand, reduction in cDNA size was observed with poliovirus RNA as template when the Mg^{2+} concentration was increased from 8 to 20 mM (51). Therefore, when using long templates, it might be advisable to keep the Mg^{2+} concentration at the level of the total dNTP concentration (52). It is noteworthy that in the presence of Mg^{2+}, the dNTP substrates have two K_m's (49). On the other hand, no significant change in the optimal Mg^{2+} concentration was observed when the dNTP concentration was varied between 10 and 400 µM (49).

Monovalent cations (K^+ or Na^+) are not required for enzymatic activity, but they can substantially increase the incorporation with certain templates. Optima for both incorporation and length must be sought for each mRNA. In some cases, omission of the monovalent cation from the reaction results in higher yields of longer

transcripts (52). Reduction of the ionic strength seems important
when long templates are transcribed.

It has been reported (30,43) that about 2 to 3 primer molecules
per template molecule are sufficient for saturation. Since high
concentrations of the primer do not inhibit the reaction, the rule
of thumb is to use the primer at a 5- to 10- fold mass excess over
the poly(A) content of the mRNA (which is usually of the order of
10%).

The yield of reverse transcription (mass synthesized cDNA/
mass mRNA template) is never 100%. The yield increases by increasing
the concentration of all four dNTPs (44) and increasing the concentration of the enzyme at saturation levels (34,43,53). It has been
reported that for globin mRNA, saturation was reached at 80 units
enzyme/µg template (43), or at a 30- to 60-fold excess of enzyme to
template molecules (53), while the four oviduct mRNAs were saturated
at 10 units/µg (34). With a fixed amount of enzyme, the amount of
cDNA product increases with increasing template input, but the
yield decreases (43).

The product of reverse transcription always contains partial
transcripts as well as full length molecules. The partial trans-
scripts are usually of discrete size when analyzed by high resolution polyacrylamide or agarose gels containing CH_3HgOH or formamide,
respectively (34,44). (Polyacrylamide gels containing urea are
never fully denaturing). A factor important for enzymatic activity
may be lost during purification (54). On the other hand, secondary
structure might be responsible for the appearance of partial products under the conditions used. This notion is supported by the
following observations: First, the nature of the limiting dNTP
affects the qualitative pattern of incomplete transcripts (44,55)
and second, a protein purified from RSV-infected chick cells (not
encoded by the viral genome) that binds to RNA and causes unwinding,
increases the percentage of full length transcripts in vitro (56).
Other parameters which affect transcript length are as follows:

a) It has been reported that 4 mM sodium pyrophosphate in
the reaction promotes the synthesis of longer products (51). It
may simply act as an RNase inhibitor.

b) Saturating amounts of enzyme (34,43,53).

c) The concentration of the dNTPs: (i) Concentrations below
50 µM are particularly inefficient (44,57) and are considered low.
(ii) If three of the dNTPs are at high concentration (100 to 1000
µM), one is at low concentration, and the amount of enzyme is non-
saturating, the pattern of incomplete transcripts shifts to higher
lengths but is intermediate to the patterns seen with all four
dNTPs at low or high concentrations (44). With saturating amounts
of enzyme, the pattern is not substantially different from that seen
when all four dNTPs are at high concentration (43). In cases where
the dNTP concentration effect was not detected (34), the enzyme
levels were saturating and concentrations below 50 µM were not tested.

(iii) Very high dNTP concentrations (greater than 2 mM) might lead
to decrease of transcript size (52).

In summary, a typical first strand synthesis for any template
could be done under the following conditions provided the optimum
concentration of monovalent cation and the saturating levels of
enzyme are known. Instead of the usual 50 mM, 100 mM Tris-HCl,
pH 8.3, should be used. This concentration is not inhibitory and
guarantees buffering of the reaction, to which enzyme storage
buffer (pH 7.0 to 7.5) will be added. The other components of the
reaction are 10 mM Mg^{2+}, a mass of primer equal to that of template,
reducing agent (30 mM β-mercaptoethanol or 10 mM DTT), 1 mM dNTPs,
trace amounts of radioactive dNTPs to follow the reaction (specific
activity is unimportant), template, monovalent cation and enzyme.

Synthesis of the Second Strand

As early as 1970, it was shown that cDNA made in the absence
of actinomycin D is partially double-stranded (58). It was further
shown that a proportion of this material rapidly regains resistance
to single-strand-specific nucleases after denaturation (59). The
presence of actinomycin D suppresses the formation of a second
strand complementary to the cDNA (60). Formation of a second
strand is also very limited, even in the absence of actinomyin D,
if the template concentration in the reverse transcription reaction
is higher than 50 µg/ml (44). On the basis of the observation that
cDNA, freed from its template by alkaline digestion, supports deoxynucleotide incorporation by new addition of RT, the existence of a
3' terminal hairpin structure was postulated (61). Preliminary
(62,63) and detailed studies (19,21,22) with globin cDNA documented
the self-priming characteristics of reverse transcripts.

Synthesis of the second strand can be accomplished by using
Pol-I or RT (Table 1) or T4 polymerase (64). When Pol-I is used,
pH conditions can be chosen (19,65) that suppress the nucleolytic
activities of the enzyme (66,67) or fragment A (68) can be employed
(69). It is important to keep the temperature at 15° (19,65,67)
and incubate for a long time (at least 4 hr) (19,65,69). An optimization study (65) also emphasizes the importance of the nature and
concentration of monovalent cation; high quality products are obtained in the presence of KCl (but not NaCl) at a concentration
between 70 and 125 mM.

An interesting modification (65,70) of the protocol, which
presumably gives higher yields by minimizing the manipulations,
is to carry out the RT and Pol-I reactions sequentially in the
same vessel, omitting the intermediate purification step. After
boiling the reverse transcription reaction mixture for 3 min,
Pol-I and components of the second reaction are added directly. It
should be noted, however, that data concerning cloning of ds-cDNA
synthesized by this method have not been presented yet.

Table 1

Characterized ds-cDNA Clones

Name of clone	mRNA template	mRNA length (-polyA) NT	Enzyme used for 2nd strand synthesis	Vector	Host	Tailing type[a] or linkers	Length of insertion[b] (BP)	Sequenced NT[c]	Ref.
	Globin								
JW101	α (human)	575	Pol-I or RT	pMB9	χ1776	A·T	383	363	96,115
pHαG1	α (human)	575	RT	pCR1	HB101	A·T		81	116
pHb72	α (rabbit)	551	Pol-I	pMB9	C600	A·T	370	All	117
pCR1α$_r$G11	α (rabbit)	551	RT	pCR1	C600	G·C	(440)		80
pCR1α$_m$G4	α (mouse)		RT	pCR1	C600	G·C	(495)		118
pHb1003	α (chicken)		Pol-I or RT	pMB9	χ1849	A·T	541	All	119
JW102	β (human)	626	Pol-I or RT	pMB9	χ1776	A·T	(569)	66	96,115
pHβG1	β (human)	626	RT	pCR1	HB101	A·T	540	51	116
pβG1	β (rabbit)	589	Pol-I	pMB9	HB101	A·T	576	All	72
pHb23	β (rabbit)	589	Pol-I	pSC101	HB101	A·T	(420)	169	120
pCR1β$_r$G19	β (rabbit)	589	RT	pCR1	C600	G·C	(543)		80
pCR1β$_m$G9	β (mouse)		RT	pCR1	C600	G·C	(540)		118
pHb1001	β (chicken)		Pol-I or RT	pMB9	χ1849	A·T	542	~490	119
JW151	γ (human)	584	Pol-I or RT	pMB9	χ1776	A·T	450	All	96,115
pHγG1	γ (human)	584	RT	pCR1	HB101	A·T	500	93	116
B52, B36, C13	(Xenopus)		Pol-I (frag. A)	pCR1	C600	G·C	(400)		69

Table 1 (contd.)

Plasmid	Gene		Polymerase	Vector	Host	Joining			Ref.
Immunoglobulin (mouse)									
pCR1-κ40	κ-light chain (MOPC-149)	∿950	Pol-I	pCR1	χ1776	A·T	700	70	121
pL21-1	κ-light chain (MOPC-21)	∿950	Pol-I	pMB9	HB101	A·T	(950)	138	121,123
p167κRI	κ-light chain (M-167)	∿950	Pol-I	pMB9	χ1776	RI linkers	(900)		124
p603αAT	α-heavy chain (M-603)	∿1550	Pol-I	pMB9	χ1776	A·T	(600)		124
pH21-1	γ1-heavy chain (MOPC-21)	∿1550	Pol-I	pMB9	χ1776	A·T	474	∿350	125
(chicken)									
pOv230	ovalbumin	1859	RT	pMB9	χ1776	A·T	1846	All	73,74
pCRlov2.1	ovalbumin	1859	Pol-I (frag. A)	pCR1	C600	G·C	1730		114
(rat)									
pAU1,pAU2	preproinsulin I		RT	pMB9	χ1776	HindIII linkers	269,182 / 339,278	354 / 347	48 / 82
pI19,pI47	preproinsulin I		Pol-I	pBR322	χ1776	G·C	755		126
pRGH1	pregrowth hormone	∿800	RT	pBR322	χ1776	HindIII linkers			102
pBR322-GH1	pregrowth hormone	∿800	Pol-I	pBR322	χ1776	A·T	(680)	All	
(human)									
pHCS-1	chorionic somatomammotropin	∿800	RT	pMB9	χ1776	RI linkers	533	All	127

Table 1 (contd.)

Name of clone	mRNA template	mRNA length (−polyA) NT	Enzyme used for 2nd strand synthesis	Vector	Host	Tailing typea or linkers	Length of in-sertion b (BP)	Se-quenced NTc	Ref.
(silkmoth, A.polyphemus)									
pAPc-401	B chorion proteins		Pol-I	pML21	HB101	A·T	565	All	103,128
pAPc-10			Pol-I	pML21	HB101	A·T	449	All	103,128
(silkmoth, B.mori)									
pBF-36	silk fibroin	16,000	Pol-I	pMB9	HB101	A·T	1420	206	129,130
pBF-39		16,000	Pol-I	pMB9	HB101	A·T	1100	62	129,130

a A·T, poly(dA)·poly(dT) tailing; G·C, oligo(dG)·oligo(dC) tailing.

b Only the net length of an insertion is considered (and not the length of tails or linkers or transcribed poly(A) sequences). Numbers in parentheses denote indirect estimates and not measurements from sequencing data or detailed restriction endonuclease analysis.

c The total number of sequenced nucleotides (in certain cases from different regions of the insertion) is indicated.

Opening of the Hairpin DNA

S1 nuclease purified from α-amylase powder from Aspergillus oryzae is a well-characterized single-strand-specific nuclease (36). It has both endo- and exonucleolytic activities and does not show sequence specificity. At a monovalent cation (NaCl) concentration of about 0.3 M, the enzyme does not show any double-stranded activity, does not nick duplexes, and does not require a critical amount of substrate. Under appropriate conditions (71), the loop of the hairpin of ds-cDNA is specifically cleaved (19,21,22), but some "nibbling" activity (71) also occurs. This might explain why clones pβG1 (72) and pOv230 (73) are each missing 13 nucleotides corresponding to the 5' end of the mRNA. In any case, calibration of the particular S1 preparation that will be used for the opening of the hairpin DNA seems important (74), although the enzyme is always used in excess.

Since S1 will remove all single-stranded material present in the ds-cDNA preparation, the final double-stranded molecules consist of 4 types: 1) Molecules representing essentially the entire mRNA sequence.   2) Molecules missing part of the region corresponding to the 5' terminal segment of the mRNA (complete second-strand copying of a partial first strand).   3) Molecules missing part of the region corresponding to the 3' terminal segment of the mRNA (incomplete second-strand copying of a complete first strand).   4) Molecules representing only the middle portion of the mRNA (incomplete second-strand copying of a partial first strand).

Construction of Hybrid Plasmids

Double-stranded cDNA can be cloned in plasmids (Table 1) or λ-vectors (75). Of the various plasmid vectors available, pBR322 (76,77) is currently the most versatile because: 1) It is small (4.3 kb) and therefore it has fewer restriction sites. In addition, the relative yield of the cloned fragment is maximized. 2) It is derived from ColE1 and therefore can be amplified, increasing the DNA yield. 3) It has five unique restriction sites that can be used for insertion of ds-cDNA. 4) It contains two selective markers, ampicillin and tetracycline resistance, that are not transposable elements. 5) It has been sequenced in its entirety (78).

To construct hybrid molecules, the vector is linearized by a restriction enzyme which cleaves only once. The ds-cDNA is inserted in this site either by the poly(dA)·poly(dT)- or oligo(dG)·oligo(dC)-tailing methods or by the use of synthetic DNA linkers containing a recognition sequence for a restriction enzyme attached to the ds-cDNA by blunt-end ligation (Table 1). In the latter case, the insertion is readily excisable. When poly(dA)·poly(dT)-tailing is used, the insertion can again be easily excised by S1 nuclease in the presence of formamide (79) provided the tail is longer than 50

base pairs. Oligo(dG)·oligo(dC)-tailing has been used to recon-
stitute certain restriction enzyme recognition sequences. For
example, elongating an EcoRI-cleaved plasmid with an oligo(dC) tail
and the ds-cDNA with an oligo(dG) tail will reconstruct an EcoRI
site (80). A Pst site can be reconstructed by adding oligo(dG)
to Pst-cleaved pBR322 and oligo(dC) to the ds-cDNA (76,81-82). In
this case between 40% (82) and 90% (81) of the ds-cDNA insertions
are excisable with Pst.

Transformation

     Although bacterial transformation had been the subject of
intense investigation since 1944 (83), attempts to transform E. coli
were unsuccessful until 1970 when it was reported that treatment
of the bacterial cells with calcium ions rendered them competent to
take up λ DNA (84). It was subsequently shown that this method could
be used to transform E. coli with naturally-occurring R-factors
(85,86). Transformation occurred with supercoiled DNA, nicked
circles, and even linear molecules indicating that ligation and
repair of the plasmid DNA can occur within the cell (87).
     Although any E. coli strain can be transformed with the cal-
cium method, the efficiency of transformation is affected by the
genotype of the host (88), the size of the inserted DNA, and the
method used for insertion. In general, plasmids containing small
insertions transform more efficiently than those with large ones.
It is necessary to eliminate the very small pieces of DNA before
inserting the ds-cDNA into the plasmids in order to enrich the
transformants for insertions close to the full length of the mRNA
template. Therefore, size selection of the ds-cDNA should precede
the construction of hybrid molecules. In certain cases, sucrose
gradient centrifugation might be sufficient, but usually polyacrylamide gel electrophoresis under nondenaturing conditions is
necessary (19). In the latter case, if tailing is chosen as the
method of joining ds-cDNA to the vector, size selection should be
done after tailing rather than before because occasionally impurities contaminating DNA extracted from gels inhibit terminal
transferase (82).
     When ds-cDNAs derived from mammalian sources are cloned, the
NIH guidelines require the use of an EK2 host-vector system. The
only certified EK2 host that can be used with plasmids is χ1776 (89).
With a modification of the Ca^{2+} procedure, transformation efficiencies between 2.5 x 10^4 and 10^5 transformants per μg of supercoiled
DNA can be obtained (90).
     Since the efficiency of transformation with hybrid molecules
is usually two to three orders of magnitude less than that with
supercoiled DNA, it was extremely difficult to obtain clones when the
ds-cDNA available was limited either because the starting material
was only available in small amounts or the mRNA of interest was
only a small proportion of the total mRNA. The modification of a

transfection procedure (91) using Mn^{2+} ion in addition to Ca^{2+} allows
10^6 transformants per μg supercoiled DNA (92). This level of transformation is equivalent to that obtained with EK1 strains of E. coli.
χ1776 must be grown and handled in completely detergent-free glassware. If these precautions are taken, between 10^3 and 10^4 transformants per μg tailed plasmid annealed to an equimolar amount of
ds-cDNA can be routinely obtained.

Recently two other transformation procedures have been reported which may increase efficiency of transformation even further.
One method (93) used Rb^+ and DMSO in addition to Ca^{2+}. This method
increased the transformation efficiency of all strains tested, but
the highest efficiency was obtained with the E. coli strain SK1590
(10^7 transformants per μg supercoiled DNA). χ1776 has not yet been
tested with this procedure. Strains are currently being developed
which will allow transformation efficiencies from one to two orders
of magnitude greater than those obtained with SK1590 (94).

The second method utilizes strains of E. coli that are temperature-sensitive in the peptidylglycan structure of the cell wall.
Transfection efficiencies of 5×10^9 transfectants per μg φXDNA
have been obtained, but transformation efficiencies are not yet
known (95). These E. coli strains are not certified as EK2 hosts.

Screening

Once a set of clones has been generated, the clones of immediate
interest must be selected (transformants containing insertions;
transformants containing insertions of a particular sequence; transformants expressing a function encoded by the inserted DNA).

Sometimes recombinant clones can be distinguished by loss of
antibiotic resistance or some other plasmid function (76). This
can be achieved by using plasmids encoding two antibiotic resistance
markers, one of which contains the site of insertion. In one case
(82), ampicillin resistance was maintained even though DNA had been
inserted into the ampicillin-resistant gene. Therefore, this method
should be used with caution.

If a highly purified mRNA has been used as template for the
synthesis of ds-cDNA, it can be used as probe, either by itself
(e.g., labeled in vitro with ^{125}I) (20), or after end-labeling with
^{32}P following mild fragmentation (96), or in the form of cDNA.
Alternatively, ^{32}P-cRNA transcribed from unlabeled cDNA by E. coli
RNA polymerase can be used (20). This method not only identifies
recombinant plasmids but also selects for the sequence of interest.

When a probe is available, the most straightforward approach
for screening is the colony hybridization method (97). In this
method, transformants are transferred to nitrocellulose filter
paper and grown. The colonies are lysed in situ, the DNA fixed to
the filter and then hybridized to the probe.

A recent modification of this method has made it possible to
rapidly screen and store large numbers of bacterial transformants

(98). Bacteria are plated directly on filters after transformation
(avoiding the time consuming and tedious task of picking colonies).
After the transformants are grown, replicas of the original filter
are made by placing a second filter on top of the first and applying pressure. While the replicas are growing, the original filters
can be screened. The plasmid DNA in the colonies can be amplified
if the filter is placed on plates containing chloramphenicol
while the colonies are still very small. With this method, several
thousand trᵤnsformants on a single 8 cm filter can be screened.
Filters with the colonies can be stored at -80° if they are grown
on plates containing glycerol and certain salts (2).

Screening is more complicated if the initial template is a
heterogeneous mixture of mRNAs. In this case, the following
methods can be used:

a) A sample of the crude mRNA preparation used for ds-cDNA
synthesis can be further purified by biochemical means so that
mRNA species are resolved and can be used as probes (see above).
However, this is rarely feasible.

b) A promising method is to synthesize cDNA of high specific
activity and generate probes by digestion of the mixture with restriction enzymes that cleave single-stranded DNA (47,48) followed
by gel electrophoresis and elution of a specific band.

c) A more tedious but powerful method is hybridization-
arrested translation (99). This method is based on the principle
that mRNA in the form of an RNA:DNA hybrid does not direct cell-
free protein synthesis. When this technique is used, there must
be a means of identifying the cell-free products either by electrophoretic profiles or by immunoprecipitation -- this method consists
of screening by using a negative result. An interesting modification which gives positive results (100) is to first hybridize
the clone in question with the mixture of mRNAs under conditions
promoting R-loop formation (101), then separate the hybrids by
agarose gel filtration and cell-free translate after melting.
Alternatively, mRNA hybridized to cloned DNA immobilized on
nitrocellulose filters can be translated following elution (102).

d) In some cases, if two or more homologous sequences must
be distinguished, secondary screening by detailed restriction
mapping might be necessary (103,104).

APPLICATIONS

Generation of Probes

By cloning ds-cDNA, specific sequences can be obtained in
amounts far in excess of what could be produced by reverse transcription alone. This amplification is particularly important when
the starting mRNA is not very abundant.

Cloned ds-cDNA has been used as a probe to identify corresponding sequences from libraries of chromosomal DNA (3), and for the restriction site mapping within and around eukaryotic structural genes (7,8), with the DNA blotting technique (105). Recombinant DNA probes containing immunoglobulin κ-light chain sequences have been used to determine the size of the corresponding transcription unit and to establish the arrangement of variable and constant regions in this unit (106). Pregrowth hormone cDNA clones were used as probes to identify a 2400 NT nuclear mRNA precursor (107).

Purification of Sequences

By its very nature, molecular cloning can be used to purify to homogeneity a particular nucleic acid sequence. Examples of sequences purified by ds-cDNA cloning are those corresponding to insulin, growth hormone and certain immunoglobulins (Table 1).

An extreme case is the purification of the approximately 100 homologous mRNA species encoding the proteins that make up the silkmoth eggshell (chorion) (103,104). Although chorion mRNAs are rather easy to purify as a group, individual species cannot be resolved by biochemical means. Clones of these sequences were identified in a primary screening by using chorion cDNA as a probe. In a secondary screening, about 20 individual sequences have been identified by detailed restriction analysis (103,104).

Similarly, from approximately 10,000 ds-cDNA clones derived from Dictyostelium mRNA, several clones representing transcription products of the 17 actin genes were identified as distinct by restriction endonuclease analysis (81).

Sequencing Studies

By using cloned ds-cDNA as material for sequencing, the complete, or almost complete, primary structure of mRNAs encoding α- and β-globin, chicken ovalbumin, rat preproinsulin, rat pregrowth hormone, etc., has been established (Table 1).

Expression

The application of ds-cDNA cloning that is most likely to lead to practical applications is the expression of eukaryotic sequences in bacteria. The first indications that eukaryotic sequences could be correctly transcribed and translated in intact prokaryotic cells came from studies concerning cloned chromosomal DNA from simple eukaryotes. In these studies, a fragment of yeast DNA complemented E. coli mutants lacking enzymatic activities (1,108).

A functional somatostatin peptide has been produced in bacteria by inserting into the β-galactosidase gene on pBR322, a chemically synthesized DNA sequence encoding the 14 amino acid residues of the hormone (109).

The synthesis and secretion of rat proinsulin by bacterial cells has been reported (82). In this study ds-cDNA copies of rat preproinsulin mRNA were inserted into the Pst site of pBR322 with the oligo(dG)·oligo(dC)-tailing method. One of the clones resulting from transformation of χ1776 expressed a fused protein bearing both insulin and penicillinase antigenic determinants. The DNA sequence of this clone shows that the insulin region is in the correct orientation and is read in phase. A stretch of six glycine residues connects the alanine, at position 182 of penicillinase, to the fourth amino acid of rat proinsulin. The amount of proinsulin detected is only a small proportion of that expected. This is probably due to proteolytic digestion of the proinsulin peptide. Similar results were obtained when the sequence for rat growth hormone (110) was inserted into the Pst site of pBR322.

In another study, the ovalbumin coding sequence was inserted (111) into the Rl site of the β-galactosidase operon in the plasmid pOp230 (112). The ovalbumin sequence was obtained by digesting the plasmid pOv230 (73,74) with the restriction endonuclease Taql. This enzyme cuts 25 base pairs to the 5' side of the initiator AUG in the ovalbumin sequence and 150 base pairs to the 3' side of the ovalbumin/pMB9 junction. A protein of 43,000 daltons which contains the antigenic determinants of ovalbumin is produced in one of the clones obtained. This protein constitutes 1.5% of the cellular protein. It has been determined (111) that ovalbumin is quite stable in extracts of E. coli. Similar results have been obtained (113) using an Hha fragment of the ovalbumin plasmid pCR1Ov2.1 (114). Hha cuts 15 base pairs to the 3' side of the initiator AUG. This fragment was inserted into the z-gene in such a way as to produce a protein in which the first five amino acids of ovalbumin are substituted by the first eight amino acids of β-galactosidase.

Acknowledgment: We thank our colleagues who substantially contributed to the updating of this review by kindly providing unpublished data.

REFERENCES

1   Carbon, J., Clarke, L., Ilgen, C. and Ratzkin, B. (1977) in Recombinant Molecules: Impact on Science and Society (Beers, R.F. and Bassett, E.G., eds.), p. 355, Raven Press, New York, NY.
2   Wensink, P., Finnegan, D., Donelson, J. and Hogness, D. (1974) Cell 3, 315-325.
3   Maniatis, T., Hardison, R.C., Lacy, E., Lauer, J., O'Connell, C., Quon, D., Sim, G.K. and Efstratiadis, A. (1978) Cell (in press).

4   Tilghman, S.M., Tiemeier, D.C., Polsky, F., Edgell, M.H.,
     Seidman, J.G., Leder, A., Enquist, L.W., Norman, B. and
     Leder, P. (1977) Proc. Nat. Acad. Sci. U.S.A. 74, 4406-4410.

5   Tonegawa, S., Brack, C., Hozumi, N. and Scholler, R. (1977)
     Proc. Nat. Acad. Sci. U.S.A. 74, 3518-3522.

6   Tilghman, S.M., Tiemeier, D.C., Seidman, J.G., Peterlin, B.M.,
     Sullivan, M., Maizel, J.V., and Leder, P. (1978) Proc. Nat.
     Acad. Sci. U.S.A. 75, 725-729.

7   Jeffreys, A.J. and Flavell, R.A. (1977) Cell 12, 1097-1108.

8   Breathnack, R., Mandel, T.L. and Chambon, P. (1977) Nature
     270, 314-319.

9   Kafatos, F.C., Efstratiadis, A., Forget, B.G. and Weissman,
     S.M. (1977) Proc. Nat. Acad. Sci. U.S.A. 74, 5618-5622.

10   Kronenberg, H., Roberts, B.E. and Efstratiadis, A. (1978),
     (submitted).

11   Gould, H.J. and Hamlyn, P.N. (1973) FEBS Lett. 30, 301-304.

12   Proudfoot, N.J. (1976) J. Mol. Biol. 107, 491-525.

13   Salser, W., Bowen, S., Browne, D., Adli, F.E., Federoff, N.,
     Fry, K., Heindell, H., Paddock, G., Poon, R., Wallace, B. and
     Whitcome, P. (1976) Fed. Proc. 35, 23-35.

14   Dayhoff, M.O. (1972) Atlas of Protein Sequence and Structures,
     Vol. 5, National Biomedical Research Foundation, Georgetown
     University, Washington, D.C.

15   Wood, K.O. and Lee, J.C. (1976) Nucl. Acids Res. 3, 1961-1971.

16   Botchan, P. (personal communication).

17   Rabbitts, T.H. (1976) Nature 260, 221-225.

18   Rougeon, F., Kourilsky, P., and Mach, B. (1975) Nucl. Acids
     Res. 2, 2365-2378.

19   Efstratiadis, A., Kafatos, F.C., Maxam, A.M. and Maniatis, T.
     (1976) Cell 7, 279-288.

20   Maniatis, T., Sim, G.K., Efstratiadis, A. and Kafatos, F.C.
     (1976) Cell 8, 163-182.

21   Higuchi, R., Paddock, G.V., Wall, R. and Salser, W. (1976)
     Proc. Nat. Acad. Sci. U.S.A. 73, 3146-3150.

22   Rougeon, F. and Mach, B. (1976) Proc. Nat. Acad. Sci. U.S.A.
     73, 3418-3422.

23   Baltimore, D. (1970) Nature 226, 1209-1211.

24   Temin, H. and Mizutani, S. (1970) Nature 226, 1211-1213.

25   Green, M. and Gerard, G.F. (1974) in Prog. Nucl. Acid Res.
     Mol. Biol. (Cohn, W.E., ed.), Vol. 14, p. 187, Academic Press,
     New York, NY.

26   Modak, M.J., Marcus, S.L. and Cavalieri, L.F. (1973) Biochem.
     Biophys. Res. Commun. 55, 1-7.

27   Loeb, L.A., Tartof, K.D., and Travaglini, E.C. (1973) Nature
     New Biol. 242, 66-69.

28   Karkas, J.D. (1974) Proc. Nat. Acad. Sci. U.S.A. 70, 3834-3838.

29   Gulati, S.C., Kacian, D.L. and Spiegelman, S. (1974) Proc. Nat.
     Acad. Sci. U.S.A. 71, 1035-1039.

30   Verma, I., Temple, G.F., Fan, H. and Baltimore, D. (1972)
     Nature New Biol. 235, 163-167.

31   Kacian, D.L., Spiegelman, S., Bank, A., Terada, M., Metafora,
      S., Dow, L. and Maks, P.A. (1972) Nature New Biol. 235,
      167-169.
32   Ross, J., Aviv, H., Scolnick, E. and Leder, P. (1972) Proc.
      Nat. Acad. Sci. U.S.A. 69, 264-268.
33   Bishop, J.O., Morton, J.G., Rosbash, M. and Richardson, M.
      (1974) Nature 250, 199-204.
34   Buell, G.N., Wickens, M.P., Farhang, P. and Schimke, R.T.
      (1978) J. Biol. Chem. 253, 2471-2482.
35   Lizardi, P.M. and Brown, D.D. (1973) Cold Spring Harbor Symp.
      Quant. Biol. 38, 701-706.
36   Efstratiadis, A. and Kafatos, F.C. (1976) in Methods in
      Molecular Biology (Last, J., ed.), Vol. 8, p. 1, Marcel Dekker,
      New York, NY.
37   Maniatis, T., Efstratiadis, A., Sim, G.K. and Kafatos, F.C.
      (1976) in Molecular Mechanisms in the Control of Gene Expression
      (Nierlich, D.P., Rutter, W.J. and Fox, C.F., eds.), Vol. 5,
      p. 513, Academic Press, New York, NY.
38   Monahan, J.J., Harris, S.E., Woo, S.L.C., Robberson, D.L. and
      O'Malley, B.W. (1976) Biochemistry 15, 223-233.
39   Verma, I.M. and Baltimore, D. (1974) in Methods in Enzymology
      (Grossman, L. and Moldave, K., eds.), Vol 29E, p. 125, Academic
      Press, New York, NY.
40   Leis, J.P. and Hurwitz, J., ibid, p. 143.
41   Kacian, D.L. and Spiegelman, S., ibid, p. 150.
42   Marcus, S.L., Modak, M.J. and Cavalieri, L.F. (1974) J. Virol.
      14, 853-859.
43   Friedman, E.Y. and Rosbash, M. (1977) Nucl. Acids Res. 4, 3455-
      3471.
44   Efstratiadis, A., Maniatis, T., Kafatos, F.C., Jeffrey, A. and
      Vournakis, J.N. (1975) Cell 4, 367-387.
45   Palmiter, R.D., Moore, P.B., Mulvihill, E.R. and Emtage, J.S.
      (1976) Cell 8, 557-572.
46   Gielen, J., Aviv, H. and Leder, P. (1974) Arch. Biochem.
      Biophys. 163, 146-154.
47   Seeburg, P.H., Shine, J., Martial, J.A., Ullrich, A., Baxter,
      J.D. and Goodman, H.M. (1977) Cell 12, 157-165.
48   Ullrich, A., Shine, J., Chirgwin, J., Pictet, R., Tischer, E.,
      Rutter, W.J. and Goodman, H.M. (1977) Science 196, 1313-1319.
49   Marcus, S.L. and Modak, M.J. (1976) Nucl. Acids Res. 3, 1473-
      1486.
50   Leis, J.P. and Hurwitz, J. (1972) J. Virol. 9, 130-142.
51   Kacian, D.L. and Myers, J. (1976) Proc. Nat. Acad. Sci. U.S.A.
      73, 2191-2195.
52   Rothenberg, E., Smotkin, D., Baltimore, D. and Weinberg, R.A.
      (1977) Nature 269, 122-126.
53   Longacre, S.S. and Rutter, W.J. (1977) J. Biol. Chem. 252, 2742-
      2752.
54   Tsiapalis, C.M. (1977) Nature 266, 27-31.

55   Devos, R., Van Emmelo, J., Celen, P., Gillis, E. and Fiers,
     W. (1977) Eur. J. Biochem. 79, 419-432.
56   Lee, S.G. and Hung, P.P. (1977) Nature 270, 366-369.
57   Weiss, G.B., Wilson, G.N., Steggles, A.W. and Anderson, W.F.
     (1976) J. Biol. Chem. 251, 3425-3431.
58   Fujinaga, K., Parsons, J.T., Beard, J.W., Beard, D. and
     Green, M. (1970) Proc. Nat. Acad. Sci. U.S.A. 67, 1432-1439.
59   Taylor, J.M., Faras, A.J., Varmus, H.E., Levinson, W.E. and
     Bishop, J.M. (1972) Biochemistry 11, 2343-2351.
60   Ruprecht, R.M., Goodman, N.C. and Spiegelman, S. (1973) Biochim.
     Biophys. Acta 294, 192-203.
61   Leis, J.P. and Hurwitz, J. (1972) Proc. Nat. Acad. Sci. U.S.A.
     69, 2331-2335.
62   Verma, I.M., Temple, G.F., Fan, H. and Baltimore, D. (1973) in
     Viral Replication and Cancer, Proc. 2nd Duran-Reynals Int.
     Symp. (Melnick, J.L., Ochoa, S. and Oro, J., eds.), p. 355,
     Barcelona.
63   Salser, W.A. (1974) Ann. Rev. Biochem. 43, 923-965.
64   Krueger, L.J., Krauss, M.R., Caryk, T.M. and Anderson, W.F.
     (1978) Biochem. Biophys. Res. Commun. 82, 60-66.
65   Wickens, M.P., Buell, G.N. and Schimke, R.T. (1978) J. Biol.
     Chem. 253, 2483-2495.
66   Lehman, I.R. and Richardson, C.C. (1964) J. Biol. Chem. 239,
     233-241.
67   Kleppe, K., Ohtsuka, E., Kleppe, R., Molineux, I. and Khorana,
     H.G. (1971) J. Mol. Biol. 56, 341-361.
68   Klenow, H., Overgaard-Hansen, K. and Patkar, S.A. (1971) Eur.
     J. Biochem. 22, 371-381.
69   Humphries, P., Old, R., Coggins, L.W., McShane, T., Watson, C.
     and Paul, J. (1978) Nucl. Acids Res. 5, 905-924.
70   Kemp, D. (personal communication).
71   Shenk, T.E., Rhodes, C., Rigby, P.W.J. and Berg, P. (1975)
     Proc. Nat. Acad. Sci. U.S.A. 72, 989-993.
72   Efstratiadis, A., Kafatos, F.C. and Maniatis, T. (1977) Cell
     10, 571-585.
73   McReynolds, L., O'Malley, B.W., Nisbet, A.D., Fothergill, J.E.,
     Givol, D., Fields, S., Robertson, M. and Brownlee, G.C. (1978)
     Nature 273, 723-728.
74   McReynolds, L.A., Catterall, J.F. and O'Malley, B.W. (1977)
     Gene 2, 217-231.
75   Blattner, F.R., Williams, B.G., Blechl, A.C., Denniston-
     Thompson, K., Faber, H.E., Furlong, L.-E., Grunwald, D.J.,
     Kiefer, D.O., Moore, D.D., Schumm, J.W., Sheldon, E.L. and
     Smithies, O. (1977) Science 196, 161-169.
76   Boyer, H.W., Betlach, M., Bolivar, F., Rodriguez, R.L.,
     Heyneker, H.L., Shine, J. and Goodman, H.M. (1977) in Re-
     combinant Molecules: Impact on Science and Society (Beers, R.F.
     and Bassett, E.G., eds), p. 9, Raven Press, New York, NY.

77 Bolivar, F., Rodriguez, R.L., Greene, P.J., Betlach, M.C., Heyneker, H.L., Boyer, H.W., Crosa, J.H. and Falkow, S. (1977) Gene 2, 95–113.

78 Sutcliff, G. (1978) (submitted).

79 Hofstetter, H., Schamböck, A., van den Berg, J. and Weissmann, C. (1976) Biochem. Biophys. Acta 454, 587–591.

80 Rougeon, F. and Mach, B. (1977) J. Biol. Chem. 252, 2209–2217.

81 Rowekamp, W. and Firtel, R. (personal communication).

82 Villa-Komaroff, L., Efstratiadis, A., Broome, S., Lomedico, P., Tizard, R., Naber, S.P., Chick, W.L. and Gilbert, W. (1978) Proc. Nat. Acad. Sci. U.S.A. (in press).

83 Avery, O.T., McLeod, C.M. and McCarty, M. (1944) J. Exp. Med. 79, 137–158.

84 Mandel, M. and Higa, A. (1970) J. Mol. Biol. 53, 159–162.

85 Cohen, S.N., Chang, A.C.Y. and Hsu, L. (1972) Proc. Nat. Acad. Sci. U.S.A. 69, 2110–2114.

86 Cohen, S.N. and Chang, A.C.Y. (1973) Proc. Nat. Acad. Sci. U.S.A. 70, 1293–1297.

87 Cohen, S.N., Chang, A.C.Y., Boyer, H.W. and Helling, R.B. (1973) Proc. Nat. Acad. Sci. U.S.A. 70, 3240–3244.

88 Collins, J. (1977) Curr. Top. Microbiol. Immunol. 78, 121.

89 Curtiss, R. III, Pereira, D.A., Hsu, J.C., Hull, S.C., Clarke, J.E., Maturin, L.F. Sr., Goldschmidt, R., Moody, R., Inouye, M. and Alexander, L. (1977) in Recombinant Molecules: Impact on Science and Society (Beers, R.F. Jr. and Bassett, E.G., eds.), p. 45, Raven Press, New York, NY.

90 Curtiss, R. III (personal communication).

91 Enea, V., Vovis, G.F. and Zinder, N.D. (1975) J. Mol. Biol. 96, 495–509.

92 Bothwell, A. (personal communication).

93 Kushner, S.R. (1978) in Proc. Int. Symp. Genetic Engineering, Elsevier/North Holland Biomedical Press, Amsterdam (in press).

94 Kushner, S. (personal communication).

95 Szalay, A. and Suzuki, M. (personal communication).

96 Wilson, J.T., Wilson, L.B., de Riel, J.K., Villa-Komaroff, L., Efstratiadis, A., Forget, B.G. and Weissman, S.M. (1978) Nucl. Acids Res. 5, 563–581.

97 Grunstein, M. and Hogness, D.S. (1975) Proc. Nat. Acad. Sci. U.S.A. 72, 3961–3965.

98 Hanahan, D. (personal communication).

99 Paterson, B.M., Roberts, B.F. and Kuff, E.L. (1977) Proc. Nat. Acad. Sci. U.S.A. 74, 4370–4374.

100 Woolford, J. and Rosbash, M. (personal communication).

101 Thomas, M., White, R.L. and Davis, R.W. (1976) Proc. Nat. Acad. Sci. U.S.A. 73, 2294–2298.

102 Harpold, M.M., Dobner, P.R., Evans, R.M. and Bancroft, F.C. (1978) Nucl. Acids Res. 5, 2039–2053.

103 Sim, G.K., Efstratiadis, A., Jones, C.W., Kafatos, F.C.,
     Kronenberg, H.M., Koehler, M., Maniatis, T., Regier, J.C.,
     Roberts, B.F. and Rosenthal, N. (1977) Cold Spring Harbor
     Symp. Quant. Biol. 42, 933–945.
104 Sim, G.K., Maniatis, T., Efstratiadis, A. and Kafatos, F.C.
     (unpublished data).
105 Southern, E.M. (1975) J. Mol. Biol. 98, 503–517.
106 Gilmore-Hebert, M., Hercules, K., Komaromy, M. and Wall, R.
     (1978) Proc. Nat. Acad. Sci. U.S.A. (in press).
107 Harpold, M.M. and Darnell, J.E. (personal communication).
108 Struhl, K., Cameron, J.R. and Davis, R.W. (1976) Proc. Nat.
     Acad. Sci. U.S.A. 73, 1471–1475.
109 Itakura, K., Hirose, T., Crea, R., Riggs, A., Heyneker, H.L.,
     Bolivar, F. and Boyer, H.W. (1977) Science 198, 1056–1063.
110 Seeburg, P. (personal communication).
111 Frasier, T. and Bruce, B. (personal communication).
112 Fuller, F. (personal communication).
113 Mercereau-Puijalon, O., Royal, A., Cami, B., Gaparin, A., Krust,
     A., Gannon, F. and Kourilsky, P. (1978) Nature 275, 505–510.
114 Humphries, P., Cochet, M., Krust, A., Gerlinger, P.,
     Kourilsky, P. and Chambon, P. (1977) Nucl. Acids Res. 4,
     2389–2406.
115 Wilson, J.T., Forget, B.G. and Weissman, S.M. (personal
     communication).
116 Little, P., Curtis, P., Coutelle, C., van den Berg, J.,
     Balgleish, R., Malcolm, S., Courtney, M., Westaway, D. and
     Williamson, R. (1978) Nature 273, 640–643.
117 Heindell, H.C., Liu, A.Y., Paddock, G.V., Studnicka, G.M.
     and Salser, W. (1978) Cell 15, 43–54.
118 Rougeon, F. and Mach, B. (1977) Gene 1, 229–239.
119 Salser, W., Liu, A.Y., Cummings, I.M., Strommer, J.,
     Padayatty, J. and Clarke, P. (1978) (submitted).
120 Browne, J.K., Paddock, G.V., Liu, A., Clarke, P., Heindell,
     H.C. and Salser, W. (1977) Science 195, 389–391.
121 Seidman, J.G., Edgell, M.H. and Leder, P. (1978) Nature 271,
     582–585.
122 Wall, R., Gilmore-Hebert, M., Higuchi, R., Komaromy, M.,
     Paddock, G.V., Strommer, J. and Salser, W. (1978) (submitted).
123 Strathearn, M.D., Strathearn, G.E., Akopiantz, P., Liu, A.Y.,
     Paddock, G.V. and Salser, W. (1978) (submitted).
124 Early, P. (personal communication).
125 Wall, R., Higuchi, R., Paddock, G.V., Toth, C., Rogers, J.,
     Clarke, P. and Salser, W. (personal communication).
126 Seeburg, P.H., Shine, J., Martial, J.A., Baxter, J.D. and
     Goodman, H.M. (1977) Nature 270, 486–494.
127 Shine, J., Seeburg, P.H., Martial, J.A., Baxter, J.D. and
     Goodman, H.M. (1977) Nature 270, 494–499.
128 Jones, C.W., Rosenthal, N., Efstratiadis, A., Sim, G.K. and
     Kafatos, F.C. (unpublished results).

129. Morrow, J.F., Wozney, J.M. and Efstratiadis, A. (1977) in Recombinant Molecules: Impact on Science and Society (Beers, R.F. Jr. and Basset, E.G., eds.), p. 409, Raven Press, New York, NY.

130. Morrow, J.F., Efstratiadis, A., Wozney, J.M. and Mucai, T. (1978) (submitted).

GENE ENRICHMENT

M.H. Edgell, S. Weaver, Nancy Haigwood and
C.A. Hutchison III

Department of Bacteriology and Immunology
Curriculum in Genetics, University of North Carolina
Chapel Hill, North Carolina 27514

The mammalian genome is large. A haploid mouse nucleus contains around 3 x 10^8 nucleotide pairs of DNA. That is 580,000 times as much as contained in the genome of the small viruses such as ϕX174 or SV40 which have been sequenced, and 1000 times as large as the E. coli genome. Were the mouse to carry genes in its DNA at the same packing density as ϕX174, it would contain more than 5 x 10^6 genes. The isolation of single genes from such a genome represents a formidable technical task. This task might be put in perspective by considering the case of enzyme purification. Even purification of the order of 5000-fold requires considerable research effort to achieve.

Conventional fractionation schemes, such as ion-exchange chromatography, may have preparative resolutions of 20- to 60-fold enrichment. That is, if the input material is spread over 200 fractions, the bulk of the molecule of interest may be found only in five of those fractions. Each of these fractions contains a number of different molecules which have been copurified on the basis of the fractionation property. Consequently, we depend on the multiplicity of usable physical differences to effect a several-step purification of the proteins. DNA fragments do not appear to have a large number of sequence-related physical properties on which to base a many-step purification scheme. DNA fragments could be exposed to affinity columns packed with various short defined DNA sequences as the binding reagent. Such columns might be capable of fractionating nonhomologous sequences, although their capabilities have not yet been developed.

Cloning techniques, on the other hand, provide very high resolution fractionation since each clone or fraction contains a

single, or very few, DNA fragments sequestered from the other fragments. Purification of a single sequence from a very large genome
using cloning techniques depends both on the ability to generate
a large number of clones and to find among this large number the
ones of interest.

Is there any need for gene enrichment? Answering this question
involves two separate issues: how many clones can reasonably be
screened, and how many clones can reasonably be generated. Techniques are available for high sensitivity screening of large numbers
of clones with the lambda cloning vectors (1). This involves
transferring some of the DNA from a plaque to a nitrocellulose
membrane for hybridization analysis and has proven to be rapid and
reliable. A single plaque may be detected containing sequences of
interest on a petri plate with around 10^4 plaques. For the plasmid
vector systems there are well-proved procedures for screening a few
thousand clones with methods that require individual handling of
isolates in order to detect those of interest (2). Recently, a new
method has been described for screening as many as 25,000 colonies
per petri plate without individual manipulation of the clones (3).
This technique involves inducing lysis of some of the cells within
the colonies and transfer of the DNA to nitrocellulose membrane
for hybridization analysis.

How many clones need to be screened? The most common case
involves fragmenting the genome with the restriction enzyme EcoRI
and generating clones from those fragments. A complex genome
(assumed to be 5×10^9 nucleotide pairs hereafter) should contain
4×10^6 fragments if the restriction sites are randomly distributed.
This seems to correspond with the number of fragments actually
found in practice (4). In order to deal with the possibility of
being statistically unlucky, one would want to screen three or four
times more clones than fragments. Consequently, to isolate a
particular EcoRI fragment from a complex genome, around 2×10^7
clones need to be scanned. With the screening methods above, about
1000 plates and filters would need to be processed. This is possible, but not convenient.

There are ways to reduce the number of clones that need to be
screened. For example, some of the NIH approved lambda vectors
will accept fragments 15 to 20 kb in length and still produce
plaques (5-7). If the complex genome were broken into random
fragments 15 kb in size, a gene present once in the genome would
be found on 1 in 3×10^5 fragments (genome size divided by fragment
size) although there will be many different but related fragments
bearing that sequence. Consequently, if those fragments were inserted into lambda by blunt-end ligation, 1.2×10^6 clones
would need to be scanned. That would require processing only 100
plates and filters. In principle, this could be reduced even
further by using a plasmid vector and larger fragments although
there are other practical difficulties with the latter vector

system. By using a partial digest to reduce the number of clones
to be screened, a larger average fragment size could be generated.

How many clones can be generated? Plasmid vectors which must
be grown in the NIH approved bacterial strain $\chi 1776$ suffer a
serious disadvantage. Despite considerable effort on the part of
many people, a reliable method of high efficiency transformation
for this strain has been difficult to attain. Transformation
efficiencies hover around 2×10^{-6} transformants per molecule
of vector DNA. On the other hand, there are several high efficiency transformation systems available for the lambda vectors
in uncrippled E. coli cells. Infection efficiencies with $CaCl_2$-
treated cells (8) or spheroplasts (9,10) generally give about
20×10^{-6} transformants per molecule of input DNA. The use of
the crippled E. coli $\chi 1776$ imposes a serious strain on ability to
carry out gene isolation from complex genomes. Restriction and
subsequent ligation with fragments gives about 5% yields or 3×10^4
transformants per µg lambda DNA. The yield per µg drops off
sharply above 1 to 2 µg input DNA per ml of reaction mix, and
therefore, to get large numbers of transformants, the volume of
the reaction must be increased. Consequently, a transformation
reaction of about 550 ml would be required to handle the EcoRI
fragments of a complex genome. Again, this is possible but not
particularly convenient. Using the larger 15 kb random fragments,
this volume would be reduced to 25 ml since fewer clones would
be needed.

The in vitro packaging system developed for recombinant DNA
by Sternberg et al. (11) appears to bring the cloning of unenriched
DNA within reach, as that system yields a transformation efficiency
of 10- to 100-fold greater than with $CaCl_2$-treated cells. With
this transformation system, several genes have been isolated from
unenriched DNA (12,13,40). This transformation system coupled
with mild gene enrichment should provide routine access to sequences within complex genomes without the necessity of working
at the high end of the system's efficiencies.

GENE ENRICHMENT METHODS

There are a few gene systems, such as the 5S and rDNA genes,
whose properties differ enough from those of their resident genomes
that the heroic applications of classical fractionation schemes
have yielded pure genes which can be cloned relatively easily (14-
17). The enrichment techniques used here, primarily isopycnic
methods, depend on the special properties of those genes and hence
are not generally transferable to other systems. Of course, if
a gene system such as that coding for histone is reiterated to a
sufficient degree, it can also be isolated from unenriched DNA (18).

Nucleic acid affinity systems have been employed in various DNA enrichment techniques (19-25). These have been hindered in their application to complex genomes by problems such as DNA degradation and network formation due to the presence of reiterated sequences within the fragment of interest. Another enrichment scheme utilizes the R-loop procedure (26) and banding in Cs_2SO_4 or CsCl which has been sufficient to bring several sequences within the range of the λ-vector/$CaCl_2$ transformation system (27,28). The large quantity of messenger RNA required for these techniques has hindered their general application to sequence enrichment and hence they have not been generally employed.

A two-stage enrichment scheme has been developed in Leder's laboratory which routinely yields a 100- to 1000-fold enrichment (29-31). The method (Figure 1) utilizes RPC5 column chromatography (32) as the first stage of enrichment. The fractions are analyzed by agarose electrophoresis followed by transfer to a nitrocellulose membrane and hybridization (33). The fractions of interest are pooled and fractionated further by high resolution preparative electrophoresis (34). The procedural details of these two techniques have been presented recently (35,36).

RPC5 ENRICHMENT

DNA restriction fragments are retained by an RPC5 column on the basis of size, A-T content, presence of single-strand ends, and one or more unknown structural features (35). The digest of a complex genome gives a broad distribution which is typically collected over 30 to 40 fractions (Figure 2). The enrichment for any given sequence depends on several properties. We have observed that different fragments may have band-widths differing considerably from each other within the same chromatography run. The enrichment also depends on where, within the distribution, the sequence of interest elutes. Defining enrichment as the fraction of the fragment recovered in the pool, the enrichments calculated for the pools estimated to contain 80% of the fragment (Figure 2) are: αB1 (fractions 3 to 4), 4; βZ1 (fractions 14 to 15), 14; and βB1 (fractions 18 to 19), 18. In these calculations the fraction of optical density (OD) in the pool is used as a measure of the fraction of the genome in the pool. This assumes that the size distribution of the fragments is the same in each fraction, which is not strictly correct. The most useful enrichment calculation for cloning purposes would be done on a number basis and not weight. The number distribution as a function of elution would be more skewed towards the earlier fractions than the OD, as can be seen from the analytical gel analysis of the RPC5 fractions (Figure 3A). That is, later fractions have fewer

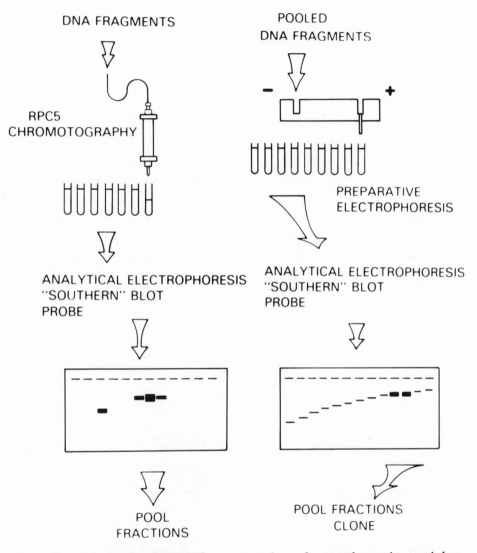

Figure 1. Schematic for RPC5 preparative electrophoresis enrichment strategy. A large quantity of the complex genome is digested
with a restriction enzyme and applied to an RPC5 chromatography
column. The fractions are analyzed by analytical gel electrophoresis,
transferred to a nitrocellulose membrane and hybridized to a high
specific activity probe to identify the interesting fractions. The
appropriate fractions are pooled and fractionated further by preparative electrophoresis. The fractions are analyzed in the same
fashion as those from RPC5 fractionation.

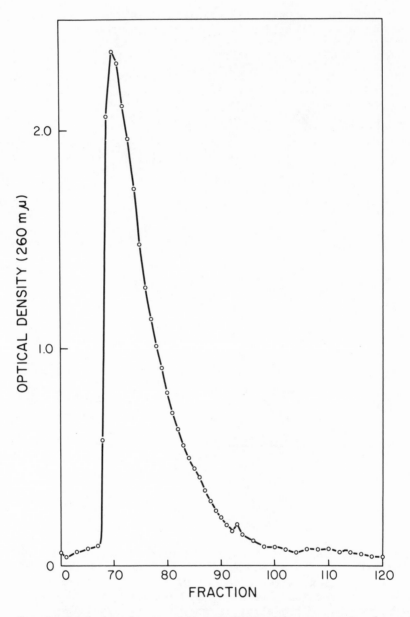

Figure 2. Optical density profile of RPC5 chromatography fractions.
Mouse DNA (50 mg C57BL/10) was digested with EcoRI, phenol extracted and fractionated with a 1.25 M to 1.8 M sodium acetate gradient
on a 2.5 x 80 cm RPC5 column. Fractions were collected and their
ODs read at 260 nm.

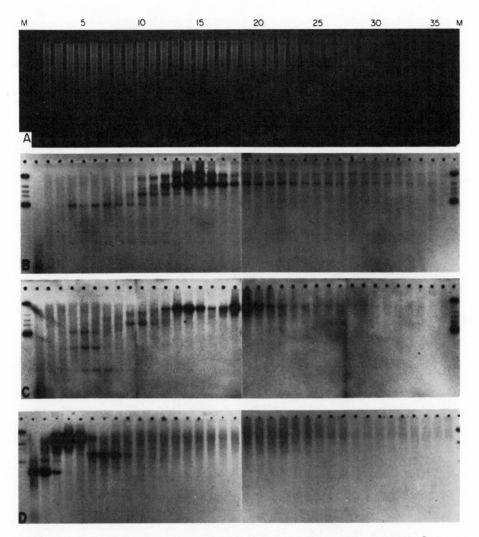

Figure 3. Analytical gel analyses of RPC5 fractions. DNA from various inbred strains of mice was digested with EcoRI and fractionated
by RPC5 chromatography. Samples of the RPC5 fractions were coprecipitated with tRNA in ethanol. The samples were electrophoresed
in submerged horizontal 1% agarose gels. The gels were run as 20 x
20 cm slabs with two rows of sample wells. The patterns have been
rearranged to display the two rows side by side rather than one
above the other as in the original gel. The smudge in lane 20 panel
is the result of overexposure of a marker channel cropped from the
figure. A. Ethidium bromide stain of C57BL DNA. B. C57BL DNA probed
with ^{32}P nick-translated α-globin sequences. C. C57BL DNA probed
with ^{32}P nick-translated α-globin sequences. D. NZB DNA probed with
^{32}P nick-translated β-globin sequences.

small fragments, and hence the OD, due mostly to the larger fragments, begins to overestimate the number of fragments.

The Southern transfer technique is not uniform with fragment
size as small fragments do not bind well to the nitrocellulose and
large fragments do not elute efficiently from the gel. So it would
be possible to miss some fragments by our analytical procedure for
detecting sequences of interest in the RPC5 fractions.

RPC5 chromatography provides the first stage of enrichment for
gene isolation and also provides a useful analytical tool for exploring eukaryotic gene organization (37). The hybridization
analyses of the RPC5 fractions have very low backgrounds which has
allowed detection of sequences that hybridize only weakly to the
probe (Figure 3B,C,D). These bands are not generally detected
when unfractionated DNA is analyzed by gel electrophoresis and
hybridization (38,39).

PREPARATIVE ELECTROPHORESIS ENRICHMENT

The second stage of enrichment involves a high resolution
preparative electrophoresis device (34). High resolution is
achieved primarily by a discontinuous collection strategy. Electrophoresis is carried out for a desired interval and then halted.
The collection chamber is drained into a fraction collector and
then refilled. The cycle is then repeated by resumption of the
electrophoresis. Usually 50% of the appropriate RPC5 pool which
contains about 1 mg of DNA is applied. The preparative electrophoresis device can accomodate large DNA loads such as 10 to 20 mg
EcoRI-digested mouse DNA. There are much less data to calculate
the theoretical enrichment of the electrophoresis due to a masking
OD which elutes with the DNA during electrophoresis. However, ethidium bromide stain can be used to gain a crude estimate of mass
distribution in the electrophoresis fractions (Figure 4). This
can be converted to a number distribution with the use of fragment
sizes. In the example shown here, a gene for an adult β-globin
was localized to a few fractions (Figure 5B) and we can estimate
the enrichment due to electrophoresis as 40-fold. Again, enrichment will depend heavily on where in the distribution a fragment
elutes. Were a 2 kb and a 20 kb fragment to elute in the same
number of fractions as the 11 kb globin-bearing fragment, they
would yield an enrichment of 13-fold and 340-fold respectively.

The combined enrichment for the β-globin fragment is calculated
to be 320-fold leaving one β-globin fragment per 15,600 fragments.
The frequency of occurrence of the β-globin fragments within the
cloned pool of fragments was actually about 1 in 2000 to 5000 (30).
This disparity between the theoretical and the actual enrichment is
due in part to the use of mass as the measure of fragment number in
the RPC5 fractions and the crudeness of the mass distribution estimates for the preparative electrophoresis fractions.

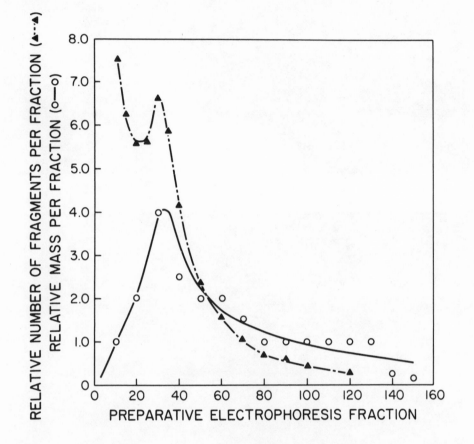

Figure 4. Preparative electrophoresis fractionation data. A crude
estimate of the mass within each fraction (———) was made by estima-
ting the amount of ethidium bromide stain in each fraction. The
size of the DNA fragments was determined from the calibration curve
generated from standards (cropped from the figure) run in the analytical gel. The relative number of fragments per fraction (---) was
calculated by dividing the estimated mass by the fragment size (in
tens of kilobases).

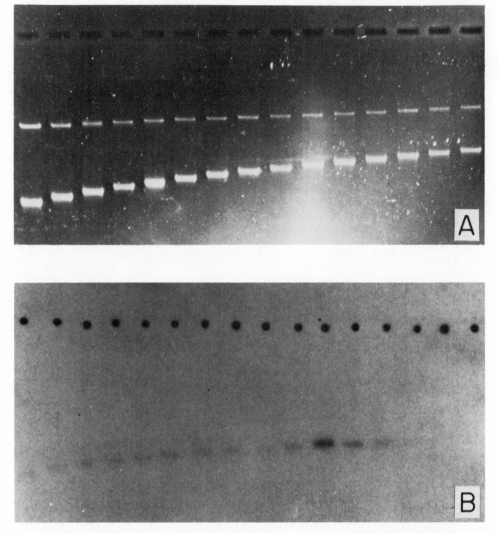

Figure 5. Analytical gel analyses of preparative electrophoresis
fractions. A pool of two RPC5 fractions containing β-globin sequences from the C57BL strain of mouse was fractionated by preparative
electrophoresis. Samples of the preparative electrophoresis fractions were coprecipitated with tRNA in ethanol. The samples were
electrophoresed as described for Figure 3. A. Ethidium bromide
stained. B. Probed with ^{32}P nick-translated β-globin sequences.
Every other fraction from the appropriate size range was analyzed.

DISADVANTAGES OF RPC5-PREPARATIVE ELECTROPHORESIS ENRICHMENT

The plastic bead support for the RPC5 chromatography system (Plaskon 2300) is no longer available commercially although some hope has been offered that alternate support may be acceptable (35). The analytic technique used to identify fractions containing sequences of interest requires very high specific activity probes (100 cpm per pg) if weakly hybridizing bands as well as major bands are to be seen. RPC5 is a fairly low resolution fractionation procedure.

Preparative electrophoresis is a slow procedure often taking 7 to 10 days for a run. After two stages of purification, particularly if there are several pools from the first, there can be a very large number of fractions to store and keep healthy.

ADVANTAGES OF RPC5-PREPARATIVE ELECTROPHORESIS ENRICHMENT

Both stages in the procedure have very high capacities coupled with very good recoveries. The DNA fragments experience very little degradation during fractionation. These techniques have proven to be highly reliable and reproducible with little demand for subtle behavior on the part of the person doing the fractionation.

SUMMARY

Isolation of specific DNA fragments from complex genomes by cloning necessitates the generation and screening of large numbers of clones for sequences of interest. These technologies are on the verge of allowing us to deal with genomes containing on the order of 10^6 DNA fragments. Mild gene enrichment can allow this isolation without the necessity of operating at the fringe efficiencies of the technology. RPC5 chromatography is a convenient, high capacity system which yields enrichments of about 5- to 20-fold. Discontinuous preparative electrophoresis is a reliable, high resolution system with high capacity capable of yielding enrichments of about 10- to 300-fold. We all look forward to the day when such enrichment will be superfluous.

Acknowledgments: The general method for gene enrichment described in this paper was the result of a team effort carried out in the laboratory of Dr. Philip Leder while M.H.E. was there on sabbatical leave. M.H.E. and C.A.H.III are supported by Public Health Service Career Development Awards. S.G.W. was supported by NIH Genetics Training Grant GM07121 and N.L.H. by Training Grant CA09057. Our research was supported by NIH grants GM21313 and AI08998.

REFERENCES

1   Benton, W.D. and Davis, R.W. (1977) Science 196, 180–182.
2   Grunstein, M. and Hogness, D.C. (1975) Proc. Nat. Acad. Sci. U.S.A. 72, 3961–3965.
3   Cami, B. and Kouritsby, P. (1978) Nucl. Acids Res. (submitted).
4   Botchan, M., Gillies, M. and Sharp, P.A. (1973) Cold Spring Harbor Symp. Quant. Biol. 38, 383–390.
5   Tiemeier, D., Enquist, L. and Leder, P. (1976) Nature 263, 526–527.
6   Leder, P., Tiemeier, D. and Enquist, L. (1977) Science 196, 1175–1177.
7   Blattner, F.R., Williams, B.G., Blechl, A.E., Thompson, K.D., Faber, H.E., Furlong, L., Grunwald, D.J., Krefer, D.O., Moore, D.D., Schumm, J.W., Sheldon, E.L. and Smithies, O. (1977) Science 196, 161–169.
8   Mandel, M. and Higa, A. (1970) J. Mol. Biol. 53, 159–162.
9   Guthrie, G.D. and Sinsheimer, R.L. (1963) Biochim. Biophys. Acta 72, 290–297.
10 Lawhorne, L., Kleber, I., Mitchell, C. and Benzinger, R. (1973) J. Virol. 12, 733–740.
11 Sternberg, M., Tiemeier, D. and Enquist, L. (1977) Gene 1, 255–280.
12 Maniatis, T., Hardison, R.C., Lacy, E., Lauer, J., O'Connell, C., Kuon, D., Sim, G.K. and Efstratiadis, A. (1978) Cell (in press).
13 Smithies, O., Blechl, A., Denniston-Thompson, K., Newell, N., Richards, J., Slightom, J., Tucker, P. and Blattner, F. (1978) Science (submitted).
14 Brown, D.D., Wensink, P.C. and Jordan, E. (1971) Proc. Nat. Acad. Sci. U.S.A. 68, 3175–3179.
15 Carroll, D. and Brown, D.D. (1976) Cell 7, 477–486.
16 Wellauer, P.K., Dawid, I.B., Brown, D.D. and Reeder, R.H. (1976) J. Mol. Biol. 105, 461–486.
17 Stafford, D.W. and Guild, W.R. (1969) Exp. Cell Res. 55, 347–350.
18 Kedes, L.H., Chang, A.C.Y., Houseman, D. and Cohen, S. (1975) Nature 255, 533–538.
19 Riggsby, W.S. (1969) Biochemistry 8, 222–230.
20 Robberson, D.L. and Davidson, N. (1972) Biochemistry 11, 533–537.
21 Shih, T.Y. and Martin, M.A. (1973) Proc. Nat. Acad. Sci. U.S.A. 70, 1697–1700.
22 Noyes, B.E. and Stark, G.R. (1975) Cell 5, 301–310.
23 Hofstetter, H., Flavell, R.A., van den Berg, J., Nierhaus, D. and Weissmann, C. (1976) Experientia 32, 797–798.
24 Woo, S.L.C., Chandra, T., Means, A.R. and O'Malley, B.W. (1977) Biochemistry 16, 5670–5676.
25 Anderson, J.N. and Schimke, R.T. (1976) Cell 7, 331–338.

26   White, R.L. and Hogness, D.S. (1977) Cell 10, 177–192.
27   Tonegawa, S., Brack, C., Hozumi, N. and Schuller, R. (1977)
     Proc. Acad. Sci. U.S.A. 74, 3518–3522.
28   Wellauer, P.K. and Dawid, I.B. (1977) Cell 10, 193–212.
29   Leder, P., Tilghman, S.M., Tiemeier, D.C., Polsky, F.I.,
     Seidman, J.G., Edgell, M.H., Enquist, L.W., Leder, A. and
     Norman, B. (1977). Cold Spring Harbor Symp. Quant. Biol.
     42, 915–920.
30   Tilghman, S.M., Tiemeier, D.C., Polsky, F., Edgell, M.H.,
     Seidman, J.G., Leder, A., Enquist, L.W., Norman, B. and Leder,
     P. (1977) Proc. Nat. Acad. Sci. U.S.A. 74, 4406–4410.
31   Seidman, J.G., Leder, A., Edgell, M.H., Polsky, F., Tilghman,
     S.M., Tiemeier, D.C. and Leder, P. (1978) Proc. Nat. Acad.
     Sci. U.S.A. (in press).
32   Hardies, S.C. and Wells, R.D. (1976) Proc. Nat. Acad. Sci.
     U.S.A. 73, 3117–3121.
33   Southern, E.M. (1975) J. Mol. Biol. 98, 503–517.
34   Polsky, F., Edgell, M.H., Seidman, J.G. and Leder, P. (1978)
     Anal. Biochem. 87, 397–410.
35   Wells, R.D., Hardies, S.C., Horn, G.T., Klein, B., Larson,
     J.E., Neuendorf, S.K., Panayotatos, N., Patient, R.K. and
     Selsing, E. (1979) Methods Enzymol. (in press).
36   Edgell, M.H. and Polsky, F. (1979) Methods Enzymol. (in press).
37   Weaver, S.G., Haigwood, N.L., Hutchison, C.A. III and
     Edgell, M.H. (1978) Proc. Nat. Acad. Sci. U.S.A. (submitted).
38   Jeffreys, A.J. and Flavell, R.A. (1977) Cell 12, 1097–1108.
39   Mears, J.G., Ramirez, F., Leibowitz, D., Nakamura, F., Bloom,
     A., Konotey-Ahulu, F. and Bank, A. (1978) Proc. Nat. Acad.
     Sci. U.S.A. 75, 1222–1226.
40   Blattner, F.R., Slightom, J., Denniston-Thompson, K., Tucker,
     P., Blechl, A., Faber, H., Richard, J. and Smithies, O. (1978)
     Science (submitted).

TRANSFORMATION OF MAMMALIAN CELLS

M. Wigler, A. Pellicer, R. Axel
Institute of Cancer Research
Department of Pathology

and

S. Silverstein
Department of Microbiology

Columbia University College of Physicians and Surgeons
701 West 168th Street
New York, New York 10032

INTRODUCTION

The introduction of foreign DNA into cells can result in a
stable and heritable change in phenotype. The ability to transfer
purified genes provides the unique opportunity to study the function and physical state of exogenous genes in the transformed host
and extends the powerful methods of virus genetics to cellular
genetics. The use of transformation[*] falls into three categories:
1) transformation as a means for gene purification; 2) transformation as a way of studying the structure and function of purified
genes; 3) transformation as a tool for dissecting complex phenotypes. This paper is concerned with the development of transforma-

---

[*]We define transformation as the introduction of foreign DNA into a
recipient cell. Transformation can frequently be detected by the
stable and heritable change in the phenotype of the recipient cell
which results from an alteration in either the biochemical or
morphologic properties in the recipient. Confusion has been generated in the past because the term was used to describe alterations
in the growth properties of cultured cells without regard to whether
these changes resulted from the acquisition of foreign DNA. In
this manuscript, we define this alteration as growth transformation
to differentiate it from the more general term transformation.

tion in cultured mammalian cells. As this is a very young field,
we shall draw mainly on our own work.

In higher eukaryotes, gene transfer has been effected by de-
bilitated virus (1) and metaphase chromosomes (2,3). Biochemical
transformation using purified DNA as donor has only recently been
effected in yeast and mammalian cells. Biochemical transformation
usually occurs at low frequencies and therefore requires the use
of appropriate selection schemes for the detection of the rare
transformant. In this review, we shall detail the development of
a successful transformation system for the biochemical transfer
of single copy eukaryotic genes utilizing total cellular DNA as
donor. The development of this system derives largely from initial
studies on the isolation and transfer of a specific DNA fragment
containing the thymidine kinase (tk) gene from the herpes simplex
virus (HSV) genome. The choice of this system was dictated by
several considerations. First, the viral genome is orders of magnitude less complex than the eukaryotic genome. This greatly
enhances the prospect for successful transformation. It allows
the possibility of purification of active restriction fragments
by size alone. Second, the tk^+ phenotype can be efficiently
selected over a tk^- phenotypic background utilizing growth conditions
in which the salvage pathway enzyme, thymidine kinase, is necessary
for survival. There exist cell lines deficient in tk with low
rates of spontaneous reversion to the tk^+ phenotype which can be
used as recipients. Third, the tk gene is an ideal subject for
mutational analysis because either the tk^+ or the tk^- phenotype can
be selected under appropriate conditions. Fourth, the gene product,
thymidine kinase, is a well-characterized viral protein of known
function that is readily distinguishable from the cellular enzyme.
Finally, growth transformation has previously been demonstrated with
viral DNA as donor (4). Successful transformation with viral genes
has provided a model system for the study of gene transfer which has
now been extended to permit the transfer of several cellular genes.

TRANSFORMATION OF tk ACTIVITY WITH FRAGMENTS OF HSV DNA

The isolation of a specific fragment of the HSV-1 genome containing the tk gene requires the identification of a restriction
endonuclease capable of digesting HSV DNA, which makes no internal
cleavages within the tk gene. Identification of such a DNA fragment using transformation as a bioassay further requires a cell
line that will stably express the tk function upon competent transformation. Ltk⁻ clone d, a clone of mouse cells resistant to
bromodeoxyuridine (BrdUrd) and deficient in cytoplasmic tk (5) was
therefore chosen for transformation experiments. Ltk⁻ cells are
unable to grow in medium containing HAT (hypoxanthine, aminopterin
and thymidine), in which survival depends upon the presence of both
salvage pathway enzymes, thymidine kinase and hypoxanthine-guanosine
phosphoribosyl transferase (6). These cells have a very low rate

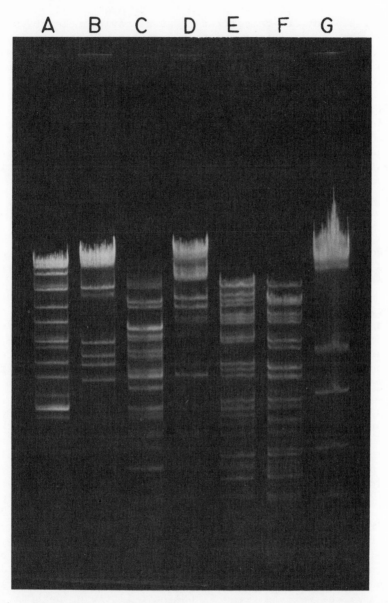

Figure 1. Digestion of HSV-1 DNA with DNA restriction endonucleases.
1.0 µg of HSV-1 DNA was incubated with 3 units of various restriction
enzymes for 3 hr at 37°C. The resultant DNA fragments were analyzed
by electrophoresis on a 17 cm 0.5% agarose slab gel. Gels were
stained with ethidium bromide and photographed under short-wave ultraviolet illumination. A) HpaI; B) BglII; C) SalI; D) HindIII; E) BamI;
F) BamI + EcoRI; G) EcoRI.

of spontaneous reversion to the tk^{+} phenotype, as judged by their
ability to form colonies in HAT-containing medium, and were used
as host recipients to demonstrate that ultraviolet-inactivated
HSV-1 virions could infect and stably confer HSV tk activity (1).

Viral DNA for transformation was extracted from virions and
cleaved with a variety of restriction endonucleases that require
the recognition of a unique hexanucleotide sequence for activity
(Figure 1). Cultures treated with DNA cleaved with BamI, SalI,
HindIII, KpnI and HpaI displayed numerous surviving colonies in
HAT. In contrast, salmon sperm DNA alone or with EcoRI-digested
HSV DNA exhibited no surviving colonies. These data suggest that
cleavage of HSV-1 DNA with each of these five enzymes generates
at least one DNA fragment containing information for the entire
tk structural gene.

                  IDENTIFICATION AND ISOLATION OF THE tk GENE

The observation that the DNA products of BamI cleavage of HSV
DNA can stably transform tk activity suggests the use of this assay
to identify the specific DNA fragment containing the tk gene. The
experimental design chosen involves the electrophoretic separation
of specific groups of DNA and ultimately of individual DNA fragments.
The fragment in which the tk gene resides is then readily identified
by transformation with these fractionated populations of DNA. To
this end, a BamI digest of HSV DNA (Figure 1) was fractionated by
electrophoresis on a 45 cm, 1% agarose slab gel. These DNA fragments were divided into five size classes and extracted from the
agarose slab.

The isolated size classes seen in Figure 2 were then used in
transformation experiments to identify the location of the tk gene.
The results of this experiment are summarized in Table 1. Transformation activity is restricted to size class III. The small
amount of activity seen in size class II probably results from the
contamination of that class with size class III, as can be seen in
Figure 2. These data indicate that the tk gene is located in one
of five well-resolved fragments ranging in size from 2.5 to 3.7 kb.
This size class was further fractionated into its five discrete
fragments (Figure 3), and the individual fragments were assayed
for their ability to transfer the tk gene (Table 2). These experiments indicate that significant transformation activity resides
only in fragment 2 of size class III. The other purified fragments
of class III as well as class II DNA have little or no activity.
The structural gene for tk is therefore contained within a single
DNA fragment 3.4 kb in length.

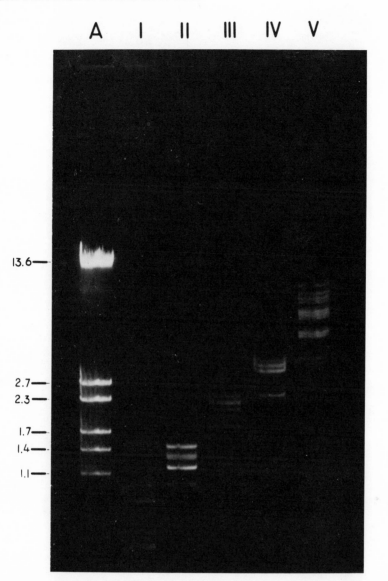

Figure 2. Fractionation of BamI-cleaved HSV-1 DNA. HSV-1 DNA was
digested with BamI endonuclease and the resultant fragments were
separated on a 45 cm 1% agarose slab gel. This preparative gel
was sliced and the DNA corresponding to five discrete size classes
was extracted from the gel and electrophoresed on a 0.5% agarose
slab. The five size classes of DNA are shown in slots I-V. Slot A
contains a preparation of EcoRI-digested Ad2 DNA as a size marker (7).

Table 1

Transformation With Size Classes of <u>Bam</u>I-Cleaved HSV-1 DNA[a]

Size class	Σ/Σ [b]
I	0/5
II	3/3
III	93/3
IV	0/5
V	0/5

[a]An equivalent number of molecules of size fractionated DNA was added to each dish.

[b]Σ/Σ, total number of colonies per number of dishes.

Table 2

Transformation With Fractionated Size Class III of <u>Bam</u>I-Cleaved HSV-1 DNA[a]

	Σ/Σ [b]	
	Experiment 1	Experiment 2
Size class II		0/2
Size class III		
Band 1		1/2
Band 2	48/4	19/2
Band 3	0/4	7/2
Band 4	0/4	0/2
Band 5	0/4	0/2

[a]An equivalent number of molecules of fractionated DNA was added to each dish.

[b]Σ/Σ, total number of colonies per number of dishes.

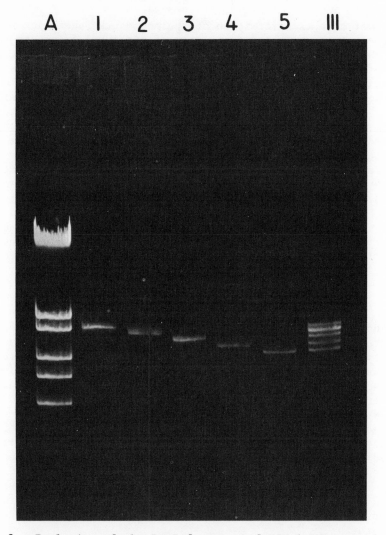

Figure 3. Isolation of the <u>BamI</u> fragment of HSV-1 DNA containing
the tk gene. The DNA bands present in size class III (Figure 2)
of a <u>BamI</u> digest of HSV-1 DNA were fractionated into five fragments
on 45 cm agarose slab gels. The isolated fragments were analyzed
by electrophoresis on a 1% agarose slab gel. Slot A contains
<u>EcoRI</u>-cleaved Ad2 DNA as molecular weight markers. Slots 1-5
contain the isolated fragments of size class III (see Figure 2).
Slot III contains the unfractionated DNA of size class III.

TRANSFORMED CELLS EXPRESS HSV tk ACTIVITY

Proof that transformation of the mouse Ltk⁻ phenotype results
from the introduction and expression of viral DNA fragments requires
a demonstration of the viral origin of the tk expressed in the
transformed clones. Although the spontaneous rate of reversion of
the recipient Ltk⁻ cells is $<10^{-9}$, it was necessary to carefully
characterize the tk activity expressed by the transformed cell
clones. The tk activity of these HAT-resistant clones is at least
20 times greater than the activity detected in the Ltk⁻ parent.
The biochemical and antigenic properties of this enzyme were
characterized by examining the neutralization of activity by specific
antisera raised against HSV-1 tk, the electrophoretic mobility of
the enzymatic activity, and the selective inhibition of tk activity
by agents specific for the viral enzyme (8). For all these parameters, the enzyme was indistinguishable from HSV-1 tk and differed
from either mouse or monkey cell tk. The conclusion that the tk
activity appearing in transfected Ltk⁻ cells results from the introduction and expression of viral DNA therefore appears firm.

PHYSICAL STATE OF THE tk GENE IN TRANSFORMED CELLS

Analysis of the stability of the transformed, tk⁺ phenotype
indicates that the transformed Ltk⁻ cells continue to express tk
activity for hundreds of generations under selective pressure.
This observation suggests that the tk gene is stably maintained in
cultured cells in a form that is recognized by the transcriptional
machinery of the host cell. It was therefore of interest to examine
the physical state of the heterologous tk gene in transformed cells.
A series of complementary experiments involving reassociation
kinetics in solution and annealings with fractionated, restriction-
cleaved cellular DNA were performed to ask the following questions.
What is the relative abundance of the tk gene in transformed cells?
Has the gene integrated covalently into the DNA of the transformants? Does integration occur at a specific locus in all transformants? Is the site of integration stable from generation
to generation?

tk Gene Frequencies by Solution Hybridization

In initial experiments, we determined the frequency of the tk
gene in a number of transformed clones by solution hybridization.
In these studies, the kinetics of reassociation of a nick-translated
tk gene were compared with the kinetics observed when the gene was
permitted to reassociate in the presence of a vast excess of transformed cell DNA. Accurate estimates of the copy number from experiments involving double-stranded DNA probes require the use of DNA
of exceedingly high specific activity. To this end, tk DNA was

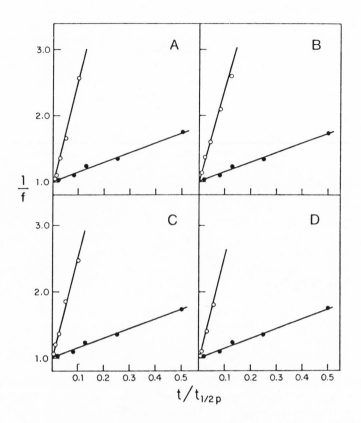

Figure 4. Renaturation of ^{32}P-labeled tk DNA in the presence of tk^{+}-
transformed clone DNA. Annealing reactions contained 0.05 ng of
^{32}P-tk DNA (specific activity 200 x 10^{6} cpm/µg) and 3 mg of DNA derived from transformed cell clones LH1b (A), LH3b (B), LH4b (C) and
LH7 (D). For comparison, tk DNA was also annealed with an equal
quantity of Ltk^{-} cellular DNA. The data are plotted according to
Sharp et al. (9), where 1/f represents the fraction of probe DNA
that is single-stranded at time t, and $t_{1/2p}$ is the time required
for half of the probe DNA to renature in the presence of untransformed cell DNA. Hybridization of tk^{+} DNA with transformed cell
DNA (o) or with Ltk^{-} DNA (•).

nick-translated with ^{32}P-deoxynucleoside triphosphates and was then
permitted to reanneal in the presence of excess quantities of cellular DNA from Ltk^{-} and tk^{+} transformed cell lines (Figure 4). In
this method of analysis, the reciprocal of the fraction of the probe
remaining single-stranded is plotted as a function of time. A linear
plot is obtained for an ideal single component, second-order reaction.

The gene frequency is determined from the ratio of the slopes observed when the probe is annealed with transformant and normal Ltk⁻ DNA. Unequal representation of different portions of the tk fragment are reflected in a deviation from linearity (9). The kinetics of annealing of the tk DNA with DNA from a variety of different clones is strikingly similar (Figure 4). For all clones, we calculate that the tk gene represents only one part in 10^6 of the mouse genome. The data therefore indicate that there probably is only one tk gene per transformed cell. No deviation from linearity is observed in Figure 4, suggesting that most of the transfecting tk fragment is equally represented in transformed cell DNA.

tk Gene Integrated in Transformed Cells

Restriction endonuclease treatment of the eukaryotic genome generates several thousand fragments which result from cleavage at precisely defined loci within the genome. The enormous complexity of eukaryotic DNA does not permit resolution of discrete fragments that contain tk gene sequences. It is possible, however, to determine the size, number and arrangement of the tk DNA fragments by eluting restriction-cleaved DNA from agarose gels onto nitrocellulose filters (10). Highly radioactive tk DNA is annealed to the DNA on these filters and the distribution of tk sequences within transformed cell DNA is then determined by autoradiography. These experiments permit us to determine the precise number of genes within a given clone, to determine whether integration of the tk gene into cellular DNA has occurred, and to compare the site of integration within cellular DNA from different clones.

A restriction endonuclease cleavage map of the purified 3.4 kb tk gene is shown in Figure 5. The enzyme EcoRI cleaves this fragment at two sites, generating three fragments 2.2, 0.68 and 0.52 kb in length. If integration has occurred, then treatment of transformed cell DNA with EcoRI should generate three fragments containing information homologous to tk DNA. A 2.2 kb fragment should result from the two internal cleavage sites. Two additional fragments are expected with molecular weights greater than 0.68 and 0.52 kb which result from a unique cleavage within the gene and cleavages at EcoRI sites in adjacent cellular DNA on both sides of the integrated fragment. From the size and number of the tk bands, additional information is obtained on the copy number and site of integration of the tk gene in each independently derived clone.

The location of the tk sequences in EcoRI-digested, transformed cell DNA is shown in Figure 6. The control in slot A includes 2 ng of total HSV-1 DNA digested with both EcoRI and BamI. As expected, the most intense band corresponds to a fragment 2.2 kb in length and reflects the internal fragment generated

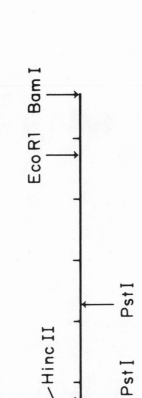

Figure 5. Restriction map of the 3.4 kb fragment of HSV DNA containing the tk gene. HSV-1 DNA was cleaved first with HpaI and then with BamI to generate a pure 3.4 kb fragment that retains transfection activity. The location of the sites for the enzymes EcoRI, HincII and PstI are shown. The fragment is calibrated in 0.5 kb units.

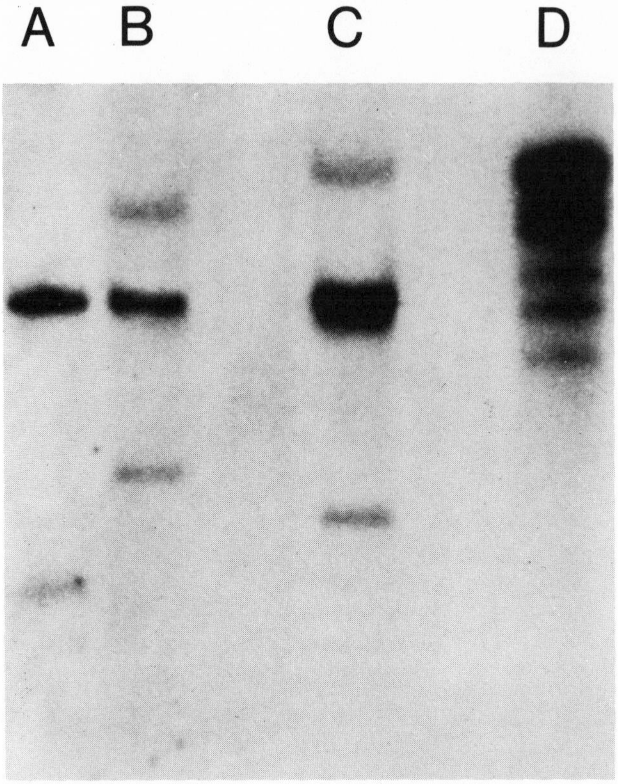

Figure 6. Identification of tk-specific sequences in DNA fragments
generated by cleavage of transformed clone DNA with EcoRI. High
molecular weight DNA obtained from clone 2b (slot B) and clone 7
(slot C) was digested with the enzyme EcoRI and electrophoresed on
0.5% agarose gels. The DNA was transferred to nitrocellulose filters and annealed with ^{32}P-tk DNA. As a control, 2 ng of HSV-1 DNA
digested with EcoRI and BamI were run in slot A. Slot D contains
^{32}P adenovirus DNA cleaved with EcoRI as markers.

upon EcoRI cleavage. Additional lower molecular weight bands are
present at a position corresponding to 0.52 and 0.68 kb fragments.
Slot B contains EcoRI-treated DNA from clone 2b, and reveals
three distinct bands 0.98, 2.2 and 3.6 kb in length. Similarly,
only three bands are observed with EcoRI-digested DNA from clone
7 -- 0.92, 2.2 and 4 kb in length.
     Utilizing this experimental approach, we examined two in-
dependently derived clones with three different enzymes -- KpnI,
HincII and EcoRI, which recognize 0, 1 and 2 sites, respectively,
within the 3.4 kb fragment containing the tk gene. We have

examined four additional clones with <u>Hinc</u>II alone. In all cases,
the number of tk fragments generated by restriction cleavage con-
forms to the predictions for the integration of maximally one tk
gene per chromosomal complement.

Analysis of the molecular weights of the fragments that contain tk sequences following restriction endonuclease digestion
of large molecular weight host DNA indicates that in all clones
studied, the tk gene is covalently integrated into cellular DNA.
In every case, the lengths of the DNA fragments generated are
larger than predicted if the tk DNA were to exist either as an
unintegrated linear or circular form. No two clones share identical
sites of integration in cellular DNA. This has been confirmed with
three different enzymes for clones 2b and 7, and with a single
enzyme for four additional clones (11). Within any given clone,
however, the site of integration must be stable and is maintained
for over fifty generations without variation in the restriction
sites flanking the gene.

We found that viral tk DNA from six independently derived
transformed clones is stably integrated into high molecular weight
host DNA. This result does not imply that integration is always
the fate of genetic elements in the transformation process in
general or even that it is always the fate of tk DNA in particular.
At present, we have no information on the potential stability of
extrachromosomal genetic elements in mammalian cells. To survive
independently as a functional entity, such an extrachromosomal
element would at the least require its own promoter for RNA transcription and an origin for DNA replication. It is not known
whether the 3.4 kb <u>Bam</u> fragment containing the structural HSV tk
gene meets either of these requirements.

TRANSFORMATION WITH SINGLE COPY GENES

The development of a system for the transfer of the HSV tk
gene to mutant mouse cells has permitted us to extend these studies
to unique cellular genes. By incorporating various improvements
into the transformation protocol, we now routinely obtain efficiencies of approximately one colony per 10^6 cells per 40 pg of
purified HSV tk gene. In the mammalian genome, a single-copy gene
is present at about one part per million. If we extrapolate from
the transformation efficiency observed for the transfer of the viral
tk gene and estimate the molecular weight of the haploid mouse
genome to be 2×10^{12} daltons, we can expect to observe the transfer
of a specific gene once per 10^6 cells per 20 µg of genomic DNA.
Under our present transformation conditions, we can therefore expect to observe transfer of single-copy genes when total genomic
DNA is used as donor.

Initial experiments designed to transfer the tk gene from
cellular DNA to mutant tk$^-$ cells were performed with donor DNA
purified from HSV tk$^+$-transformed Ltk$^-$ mouse cells. The choice of

Table 3

Transformation Data: HSV tk Gene[a]

DNA source	Σ/Σ [b]
Ltk$^-$	0/20
LH7	95/10
LH2b	16/9
LHHB	78/10
LHH5-1	4/20

[a] 20 μg of high molecular weight DNA was added to each dish as described in Wigler et al. (12).

[b] Σ/Σ, total number of colonies per number of dishes.

this donor for initial studies was dictated by several considerations First, we have previously shown that HSV tk$^+$ cells contain only a single copy of the viral gene per cellular genome (11). Second, the properties of the viral enzyme are sufficiently different from those of the murine enzyme to allow characterization of the acquired tk activity by gel electrophoresis. Finally, the availability of purified restriction fragments containing the viral tk gene allows us to detect and analyze the physical state of the transferred gene in the DNA of the transformant.

Transformation was attempted using DNA purified from four independently derived clones of Ltk$^-$ containing the viral tk gene. Transformation assays with DNA purified from the four HSV tk$^+$ transformants gave rise to numerous colonies (Table 3). As expected, DNA obtained from Ltk$^-$ was unable to transfer tk activity to Ltk$^-$ cells. For clarity, we define primary transformants as the original HSV tk$^+$ mouse cells derived following transfer of purified viral DNA. We define secondary transformants as tk$^+$ cells obtained following transfer of cellular DNA extracted from primary transformants. It is apparent from Table 3 that the frequency of transformation varies for DNA derived from different sources. DNA derived from clones of LH2b, LH7 and LHHB resulted in transformation frequencies 3 to 16 times greater than predicted. DNA from clone LHH5-1 generated colonies at a frequency less than that predicted above.

ORIGIN OF tk ACTIVITY IN SECONDARY TRANSFORMANTS

The transformation frequencies observed (Table 3) range from one colony per 1×10^5 cells to one colony per 5×10^6 cells. In studies with the recipient cell Ltk$^-$ over the past years, we have never observed a single spontaneous revertant. Our estimate of the rate of spontaneous reversion of Ltk$^-$ to tk$^+$ is $<10^{-9}$. The appearance of even a single colony in cellular transformation experiments is therefore significant, and strongly suggests that expression of the tk$^+$ phenotype results from the introduction and expression of foreign DNA. Nevertheless, the expression of tk activity in these transformed cells could conceivably result from either reversion or reactivation of wild-type enzyme rather than the introduction and expression of a new tk gene from donor DNA. Analysis of the electrophoretic properties of the tk activities of the transformed cells allows us to distinguish among these possibilities. The size and charge of the murine and viral tk activities are sufficiently different to permit separation by nondenaturing polyacrylamide gel electrophoresis (13).

The migration of the tk activity from four independently derived secondary transformants is indistinguishable from that of the viral enzyme and readily separable from murine tk activity. Furthermore, the use of donor DNA derived from cells originally transformed with viral tk DNA allows a direct analysis of the physical state of the tk gene in recipient cells. Therefore, the filter hybridization technique discussed earlier is utilized to identify the number and location of the tk sequences liberated upon restriction endonuclease cleavage of transformed cell DNA. The data indicate that the structural gene for tk is present in both primary and secondary transformants (11). The identification of viral tk activity and the detection of HSV tk gene sequences in the DNA of transformed Ltk$^-$ cells demonstrated that the transformation observed using total cellular DNA as donor results from the introduction and expression of DNA sequences coding for the viral tk.

TRANSFORMATION WITH INDIGENOUS CELLULAR GENES

These experiments have demonstrated the feasibility of transferring a unique gene without prior fractionation of the donor genome. We therefore attempted the transfer of indigenous cellular genes. High molecular weight DNA was isolated from LM, a line of mouse cells that expresses tk activity, and also from mouse liver. Transformation was carried out as described earlier, and after two weeks, colonies surviving in HAT medium were scored. With LM DNA, 65 colonies were observed in 10 culture dishes, and 28 colonies were observed in four culture dishes with mouse liver DNA (Table 4). In contrast, Ltk$^-$ DNA failed to produce a single colony.

Table 4

Transformation Data: Indigenous tk Gene[a]

DNA source	Σ/Σ [b]
Ltk⁻ (mouse cells)	0/30
Drosophila embryo cells	0/10
Slime mold	0/10
Salmon sperm	0/10
LM (mouse cells)	63/10
Mouse liver	28/4
CHO (hamster cells)	72/10
Chicken RBC	31/10
Calf thymus	62/8
HeLa (human cells)	9/9

[a]20 µg of high molecular weight DNA was added to each dish as described in Wigler et al. (12).

[b]Σ/Σ, total number of colonies per number of dishes.

These experiments demonstrated the feasibility of intraspecific gene transfer. We next asked whether transformation could also be effected with DNA from distantly related eukaryotic organisms. High molecular weight DNA was purified from Dictyostelium, Drosophila embryo cultures, salmon sperm, chick erythrocytes, cultured hamster cells, calf thymus and HeLa cells. Chick, calf, hamster and human DNA generated numerous surviving colonies, while no transformation was observed with Dictyostelium, Drosophila or salmon DNA. We conclude that both intra- and interspecific transfer of the tk gene can be effected with high efficiency under our transformation conditions.

tk ACTIVITY OF TRANSFORMANTS IS DONOR DERIVED

The appearance of surviving colonies following transformation assays with cellular DNA could result from reactivation of the murine tk or from the introduction of a new wild-type tk gene coded for by donor DNA. As discussed earlier, the exceedingly low frequency of spontaneous reversion of the recipient cells, coupled with the inability to generate tk⁺ transformants using Ltk⁻ DNA as donor, argues strongly that the tk⁺ phenotype observed following transformation results from the introduction of a new structural tk

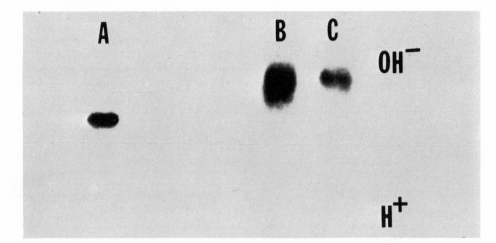

Figure 7. Isoelectric focusing of thymidine kinase activity in gels.
The 30,000 x g supernatants from homogenates of LM (a wild-type
mouse cell (slot A)), HeLa (a human cell line (slot B)) and L(HeLa)-
1 (a tk⁻ mouse cell transformed using DNA from tk⁺ HeLa cells (slot
C)) were focused on 4.5% acrylamide gels. Thymidine kinase activity
was assayed in situ, and the product was blotted out onto PEI-
cellulose and localized by fluorography as described (12).

gene into tk⁻ cells. The human tk enzyme displays biochemical properties distinct from those of the mouse, enabling us to determine
the source of the tk expressed in transformants. The pI of human
tk is 9.7, whereas the murine tk activity has a pI of 9.0 (13).
Extracts of LM cells, HeLa cells and transformants generated with
purified HeLa DNA were analyzed by isoelectric focusing in polyacrylamide slabs. The tk activity was localized by assaying the
conversion of dThd to dTMP in situ. The product of this reaction,
³H-dTMP, was blotted out of the gel onto PEI plates which were then
analyzed by fluorography. Figure 7 demonstrates that the pI of
transformed cell tk is identical to that of human tk and differs
from the more acidic murine tk. Transformation must therefore result from the expression of the donor tk gene.

GENERALITY OF THE TRANSFORMATION PROCESS

The method used to transfer the thymidine kinase gene can, in
principle, be applied to any gene for which conditional selection
criteria are available. Thus far, we have restricted this discussion
to the thymidine kinase gene. To explore the generality of the
transformation process, we next asked whether DNA-mediated transfer
could be effected to complement APRT⁻ cells. The major salvage
pathway for adenosine biosynthesis in vertebrates involves the enzyme

Table 5

Transformation Data: Indigenous tk and APRT Genes[a]

DNA source	Total no. colonies tk+ per total no. dishes	Total no. colonies APRT+ per total no. dishes
Hamster (CHO cells)	22/5	20/14
Human (HeLa cells)	42/4	95/14
Mouse (LH2b cells)	100/5	24/15
Salmon (testes)	0/5	0/15

[a] 20 μg of high molecular weight DNA was added to each dish using the procedure described in Wigler et al. (12).

adenine phosphoribosyltransferase (APRT) which converts adenine to AMP. Mutant cells deficient in APRT, cultured in the presence of azaserine (an inhibitor of the de novo pathway for purine bio-synthesis) and adenine cannot survive, whereas wild-type cells can. Therefore, growth in this selective medium was used as a bioassay for the transfer of the APRT gene to APRT⁻ mouse L cells using donor DNA from a variety of vertebrates. The results of the transformation assay are shown in Table 5. DNA from human, hamster, mouse and rabbit all generate numerous surviving colonies when APRT⁻ cells are grown in the selective medium. DNA from APRT⁻ cells is unable to effect successful transformation.

It is necessary to demonstrate that the APRT activity expressed by transformants results from the introduction of a heterologous wild-type APRT gene via DNA-mediated gene transfer and not from reactivation of recipient cell APRT. To this end, we exploited the differences in pI between donor and recipient cells' APRT activity. The APRT activity of donor and recipient cells and transformants was analyzed by isoelectric focusing (Figure 8). In all instances, the APRT activity of the transformants appeared to be donor derived (Wigler et al., submitted for publication).

Detection of gene transfer for the recessive markers, tk and APRT, required the use of appropriate mutant cell lines. The ability to transfer dominant markers is not restricted to specific mutant cells and would greatly extend the usefulness of the trans-formation technology. Wild-type folate reductase from mammalian cells is inhibited by methotrexate (mtx). Structural mutants for this enzyme have been obtained that are resistant to high concentrations of this drug (14). In preliminary experiments, we have transferred this mtx resistant folate reductase to mtx sensitive cells (Table 6).

A B C D E   F   G H   I J   K L M   NO

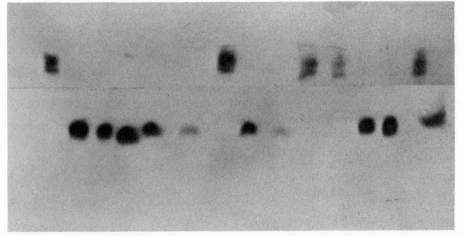

Figure 8.   Isoelectric focusing of adenine-phosphoribose transferase
in gels.   The high speed supernatants from homogenates of wild-type
cells, tissue and transformants were focused on 4.5% acrylamide gels
containing an Ampholine mixture of 0.8%, pH 2.5-4, 0.8%, pH 4-6, and
0.4%, pH 5-7.   For development of enzyme activity, 2-^{3}H-adenine was
used and the product was blotted out onto PEI-cellulose and localized
by fluorography.   A)Ltk^{-} cell extract; B) rabbit liver homogenate;
extracts from C) Hep-2, D) CHO, E) Ltk^{-}, APRT^{-} cells transformed with
HeLa cell DNA, F) cells transformed with CHO cell DNA,G) cells transformed with LH2b cell DNA, H) cells transformed with HeLa cell DNA,
I) cells transformed with CHO cell DNA, J) cells transformed with LH2b
cell DNA, K) an Ltk^{-}, APRT^{-} revertant, L) HEp-2 cells, M) CHO cells,
N) Ltk^{-} cells, O) rabbit liver homogenate.

Table 6

Transformation Data: Mtx^{Ra}

DNA source	Total no. colonies mtx^R per total no. dishes	Total no. colonies tk$^+$ per total no. dishes
CHO cells A29^b	56/5	25/5
CHO cells wild-type	0/5	30/5

[a]Ltk$^-$, APRT$^-$ cells were used as recipients. 20 µg of DNA was used to transform 10^6 cells/dish and 24 hr after DNA was added to the dish, medium containing 0.2 µg/ml of methotrexate was added.
[b]CHO Pro$^-$Mtx^{RIII}, Flintoff et al. (14).

SUMMARY

The transfer of specific genes, free of chromosomal protein, may facilitate the analysis of the control of gene expression in complex eukaryotes. The availability of sensitive assay systems for transformation may ultimately allow the isolation of any gene for which selective growth conditions exist. To explore this possibility, we developed a transformation system for the thymidine kinase (tk) gene of herpes simplex virus, HSV-1. Through a series of electrophoretic fractionations in concert with transformation assays, we isolated a unique 3.4 kb fragment of viral DNA capable of efficiently transferring tk activity to mutant Ltk$^-$ cells (8). Analysis of the transformed cell DNA in molecular hybridization experiments demonstrated that a single copy of the tk gene was covalently integrated into the DNA of all transformants (11).

The development of a system for the transfer of the HSV tk gene to mutant mouse cells has permitted us to extend these studies to unique cellular genes. In addition, the availability of cell lines bearing a single copy of the HSV tk gene has allowed us to trace the fate of this gene when DNA from these cells is used as donor in transformation experiments. We have found that high molecular weight DNA obtained from tk$^+$ tissues and cultured cells from a variety of organisms can be used to transfer tk activity to tk$^-$ mutant mouse cells. The resulting tk activity expressed in recipient cells is donor derived (12).

The generality of the transformation process has been demonstrated by experiments in which we have successfully transferred the APRT gene and an mtx resistance folate reductase gene using total cellular DNA as donor. The method used to transfer these genes can, in principle, be applied to any gene for which conditional selection criteria are available. In practice, the efficiency of

gene transfer can be expected to be a function of the recipient
cell, the source of the gene being transferred and the stringency
of the selection criteria. In order for gene transfer to be
readily detectable, it must occur at a frequency higher than the
spontaneous rate of mutation of the recipient to the phenotype
selected. The frequencies observed for the transfer of the tk
gene to Ltk$^-$ range from 2×10^{-7} to 1×10^{-5}. This is also the
frequency range observed for spontaneous mutation at many interesting loci in cultured somatic cells. Improvements in transformation efficiency or prefractionation of donor DNA can be
expected to extend the usefulness of this technique.

Transfer of single copy genes in eukaryotes has also been
achieved using metaphase chromosomes as donor (2,3). The transfer
of single copy genes using genomic DNA as donor has advantages:
DNA can be obtained from interphase cells; genomic DNA can be
cleaved with restriction enzymes and subsequently fractionated;
distances between linked genes can, in theory, be precisely
determined; and, most important, DNA-mediated gene transfer can
be used as a bioassay allowing the purification and subsequent
amplification of specific genes.

Transformation with restriction endonuclease-cleaved, size-
fractionated viral DNA fragments has allowed the purification
of viral genes responsible for growth transformation (4) and the
herpes simplex genes coding for thymidine kinase (8,15). This
approach, while successful with viral genomes, cannot be used
to purify the single copy genes of the vastly more complex
mammalian genomes. To purify such genes, for which specific
hybridization probes are unavailable, may require the construction of recombinant molecules and transformation into both prokaryotic and eukaryotic cells.

REFERENCES

1   Munyon, W., Kraiselburd, E., Davies, D. and Mann, J. (1971)
    J. Virol. 7, 813-820.
2   McBride, O.W. and Ozer, H.L. (1973) Proc. Nat. Acad. Sci.
    U.S.A. 70, 1258-1262.
3   Willecke, K. and Ruddle, F.H. (1975) Proc. Nat. Acad. Sci.
    U.S.A. 72, 1792-1796.
4   Graham, F.L., Abrahams, P.J., Mulder, C., Heijneker, H.L.,
    Warnaar, S.O., de Vries, F.A.J., Fiers, W. and van der Eb,
    A.J. (1974) Cold Spring Harbor Symp. Quant Biol. 39, 637-650.
5   Kit, S., Dubbs, D., Piekarski, L. and Hsu, T. (1963) Exp.
    Cell Res. 31, 297-312.
6   Littlefield, J. (1974) Science 145, 709-710.
7   Pettersson, U., Mulder, C., Delius, H. and Sharp, P.A. (1973)
    Proc. Nat. Acad. Sci. U.S.A. 70, 200-204.
8   Wigler, M., Silverstein, S., Lee, L.-S., Pellicer, A., Cheng,
    Y.-c. and Axel, R. (1977) Cell 11, 223-232.

9   Sharp, P.A., Pettersson, U. and Sambrook, J. (1974) J. Mol.
     Biol. 86, 709-726.
10   Southern, E.M. (1975) J. Mol. Biol. 98, 503-517.
11   Pellicer, A., Wigler, M., Axel, R. and Silverstein, S. (1978)
     Cell 14, 133-141.
12   Wigler, M., Pellicer, A., Silverstein, S. and Axel, R. (1978)
     Cell 14, 725-731.
13   Kit, S., Leung, W.C., Trkula, D. and Jorgensen, G. (1974)
     Int. J. Cancer 13, 203-218.
14   Flintoff, W.F., Davidson, S.V. and Siminovitch, L. (1976)
     Somatic Cell Genet. 2, 245-261.

CONSTRUCTED MUTANTS OF SIMIAN VIRUS 40

D. Shortle, J. Pipas, Sondra Lazarowitz, D. DiMaio
and D. Nathans

Department of Microbiology
Johns Hopkins University School of Medicine
Baltimore, Maryland   21205

Genetic analysis of cells and viruses is based on mutants with
specific physiological defects. Classically, such mutants have
been generated by random mutagenesis followed by selection of a
desired phenotype, an approach resembling spontaneous mutation and
selection. Recent advances in nucleic acid biochemistry have led
to a new, more active method of genetic analysis whereby viral
genomes with defined alterations at preselected sites are constructed
in vitro and mutant viruses are subsequently cloned from the progeny
of such modified genomes. By this means, one can perturb specific
sites in a DNA molecule, within a coding sequence or a regulatory
element, and determine the effect on function of the genome or its
products without the necessity of phenotype selection. In most
instances, site specificity is based on the recognition and cleavage
of DNA sequences by restriction endonucleases, and alterations of
the genome are carried out by enzymatic and/or chemical modifications
at or near restriction sites. Thus, once a detailed cleavage map
of a given DNA molecule is available, and especially if the nucleotide sequence is known, mutants can be isolated with specific base
changes, deletions, substitutions or rearrangements in any part of
the molecule.
      In this chapter, we describe the construction, cloning and
identification of deletion mutants and point mutants of Simian
Virus 40 (SV40), a small tumorigenic papovavirus whose genome is a
covalently closed, circular DNA duplex of 5226 nucleotide pairs.
Because of the early application of the new methods for analyzing
and restructuring DNA to the genome of SV40, this virus has served
as a model for the in vitro construction of mutants. The same
methods, however, are applicable to any DNA molecule that can replicate within cells. Therefore, though we concentrate on constructed

mutants of SV40 in this review, the general approaches, and often
the specific procedures, can be used to construct mutants of plasmids or of other viruses, including those containing segments of
cellular or synthetic DNA.

GENERAL SCHEME FOR CONSTRUCTING, ISOLATING
AND IDENTIFYING MUTANTS OF SV40

     The general scheme for generating constructed mutants of SV40
involves the following steps: 1) site selection, generally by
cleavage of viral DNA with restriction endonucleases; 2) local
enzymatic and/or chemical modifications of the DNA; 3) cloning by
transfection of cell monolayers with the suitably modified DNA,
and 4) identification of a cloned mutant by mapping the mutational
site. Each of these steps will be described in the sections below,
following which a series of sample protocols, now in use in our
laboratory, will be presented.

SITE SELECTION

     As already noted, selection of sites or regions of a genome
to be altered is generally based on the cleavage specificity of
restriction endonucleases. In the case of SV40 DNA, a detailed
cleavage map has been constructed (Figure 1) and the entire nucleotide sequence is known (1,2). The simplest site selection is by
use of a restriction enzyme that recognizes only one site in SV40
DNA. A molecule opened at that site by double-strand cleavage
serves as the starting point for isolation of deletion mutants
lacking DNA segments surrounding the original restriction site (3,
4). Covalently closed circular DNA duplex (form I DNA) may also
be nicked (i.e., cleaved in only one strand) at the restriction
site, the nicked molecules serving as the starting material for
local point mutagenesis (see below). Multicut restriction enzymes
can be used to construct deletion or point mutants at a preselected
site, also. In the presence of appropriate concentrations of
ethidium bromide and a multicut enzyme, cleavage of form I DNA is
inhibited either after one single-strand scission per molecule or
after one double-strand break, due to enhanced binding of the dye
by open circular (form II) or linear DNA (5). The optimal concentration of ethidium bromide must be determined for each restriction
enzyme; however, with some enzymes intermediate nicked molecules
have not been observed at any ethidium bromide concentration. Multicut restriction enzymes, or combinations of two different enzymes,
can also be used to excise a defined segment of DNA from the genome
(3,6). For example, a multicut enzyme may first be used to open
SV40 form I in the presence of ethidium bromide (or by partial digestion), and the isolated full length linear DNA can be subsequently
cleaved with a single cut enzyme (7). The linear products of desired

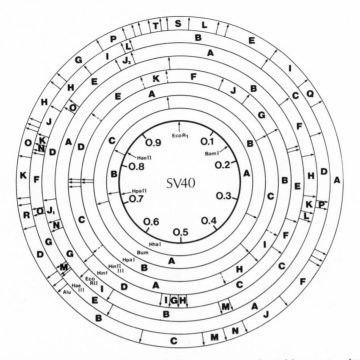

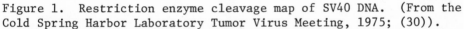

Figure 1.   Restriction enzyme cleavage map of SV40 DNA. (From the
Cold Spring Harbor Laboratory Tumor Virus Meeting, 1975; (30)).

length are then isolated by gel electrophoresis. Thus, by varying
conditions and enzymes, circular molecules containing a single
nick at a defined site (or at one of several enzyme sites), or
linear DNA with defined ends can be isolated (Figure 2).
      Two recent extensions of these methods should broaden the range
of site selection considerably. In the first procedure, nicks at
a specific restriction site are "translated" (i.e., moved) by controlled DNA polymerase action to an adjacent region and thus serve
as the starting point for construction of small deletion mutants
or point mutants within that region (DiMaio, unpublished results;
Figure 3). In a second, more general method (diagrammed in Figure
4), a single, randomly positioned gap is introduced into each DNA
molecule by first nicking with pancreatic DNase (or a multicut
restriction enzyme) in the presence of ethidium bromide, followed
by limited exonuclease digestion at the nick. The gapped molecules
are then annealed to an immobilized specific SV40 restriction fragment derived from the genomic region of interest. The subpopulation
of molecules that anneal to the restriction fragment contain gaps
within the corresponding genome segment and can therefore be used
to construct local point mutants or deletion mutants (Shortle,
unpublished results).

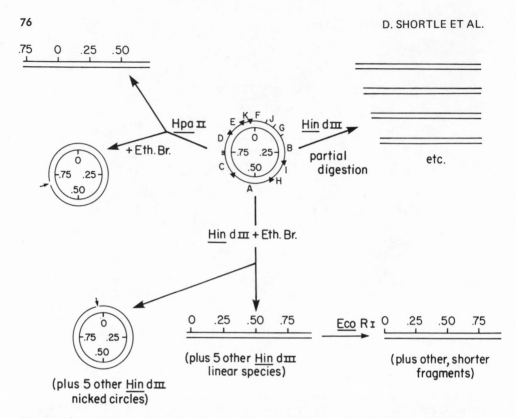

Figure 2. Site selection by double-strand or single-strand scission of SV40 DNA with restriction endonucleases. F, J, G, etc. refer to HindII and III restriction fragments. Triangles indicate HindIII restriction sites.

Mention should be made here of the construction of deletion mutants of SV40 by random opening of form I DNA, e.g., by pancreatic DNase in the presence of Mn^{2+} (8), or random nicking in the presence of ethidium bromide followed by S1 nuclease (9). Although these procedures do not involve preselection of sites to be mutagenized, subsequent cloning of individual mutants and restriction analysis of mutant DNA readily identifies the site of deletion.

GENERATION OF DELETION MUTANTS BY CYCLIZATION OF LINEAR DNA

After linear SV40 DNA has been prepared by one of the procedures outlined above, circular viral genomes missing DNA sequences from the ends of the linear molecule can be generated in one of two ways: 1) by enzymatic cyclization in vitro with DNA ligase (10,11), or 2) by intracellular cyclization following transfection (3). In either case, full length linear molecules are first treated with

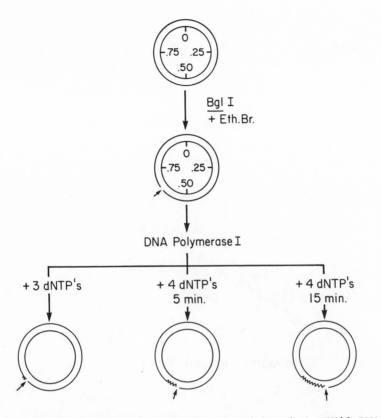

Figure 3. Repositioning of a single-strand break in SV40 DNA by controlled nick translation with E. coli DNA polymerase I.

exonuclease and/or S1 nuclease prior to cyclization or transfection in order to remove nucleotides at the ends thereby markedly reducing the regeneration of wild-type DNA (4,9). In vitro cyclization of blunt-ended (or S1-treated) DNA can be effected with T4 ligase, which does not require cohesive termini (12). The circular product can then be used directly for transfection or, preferably, first purified by agarose gel electrophoresis or by selective digestion with ATP-dependent DNase (13) to eliminate residual linear molecules or undesired ligation products.

Intracellular cyclization of short linear SV40 DNA molecules can occur via cohesive termini (from linear DNA prepared by excision of a segment with an appropriate restriction enzyme (3,6,14)), or more generally, by a mechanistically obscure reaction in which the ends of the linear molecule are covalently joined with loss of a variable number of nucleotides at each end (3). Thus, any linear molecule will generate a variety of extended deletion mutants missing from a few to hundreds of nucleotide pairs. Nucleotide sequences at the deletion joints of a number of such extended

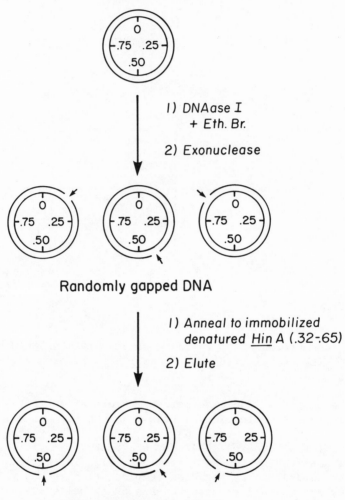

Figure 4. Fractionation of randomly gapped DNA by
annealing to immobilized restriction fragments.

deletion mutants have been determined (M. Gutai, personal communication). In most instances, the joint sequence is that expected
for a simple deletion: only sequences derived from near the ends of
the initial linear molecule are found. The simplest way such a
joint could arise is by enzymatic trimming at the ends of the transfecting linear DNA and subsequent blunt-end ligation, although a
more complex recombination mechanism is certainly possible. In
some instances, the joint has a new stretch of nucleotides joining
the parental sequences; the origin of such small insertions is
not known.

GENERATION OF BASE SUBSTITUTIONS BY LOCAL MUTAGENESIS

Generation of base substitutions by the recently described
local mutagenesis method (15) starts with specifically nicked
viral DNA, prepared as described earlier (see Figures 2 and 3). In
the simplest case, an enzyme that recognizes only one site in the
molecule is used, for example HpaII nicking of SV40 DNA I, as
illustrated in Figure 5. The nick is then extended into a small
gap by limited exonucleolytic digestion with DNA polymerase I or
exonuclease III. With Pol-I from M. luteus, the gap can be limited
to a length of about 5 nucleotides extending 5' from the original
nick. It should be noted that only one strand in any given molecule is nicked, and that different nucleotides are exposed in each
of the two DNA strands. The gapped molecules are then treated with
the single-strand specific mutagen, sodium bisulfite, which deamin-
ates cytosine to uracil (16,17). Such mutagenized, gapped molecules

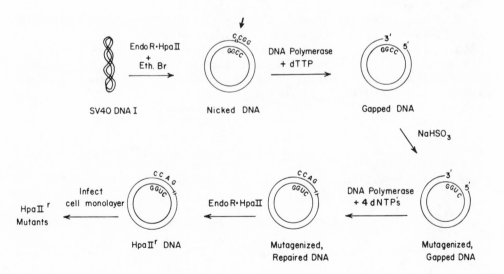

Figure 5.   Base substitution by local mutagenesis
            with sodium bisulfite (15).

can be used directly for mutant isolation or, where a restriction enzyme that recognizes only one site in the DNA molecule has been used to select the site of a gap, the mutagenized gap can be filled in with DNA polymerase in the presence of all four deoxynucleoside triphosphates and molecules resistant to the one-cut enzyme purified prior to isolation of mutants.

Several features of the local mutagenesis method are worth noting. First, with bisulfite as the mutagen, deamination of a cytosine residue within the gap results in an unambiguous change in base-pairing properties since the natural base, uracil, is produced. Moreover, because the uracil is in a single-stranded segment of DNA, gap-filling in vitro or within the cell results in the substitution of an A-U pair for the original G-C pair. Thus, the. mutation cannot be reversed by cellular repair enzymes. Second, the extent of deamination of exposed cytosine residues can be controlled by the reaction time and/or bisulfite concentration. Even under conditions that deaminate over 60% of cytosine residues in single-stranded DNA, there is essentially no effect on duplex DNA as assessed by specific infectivity of bisulfite-treated SV40 DNA form II (15). Third, the size of the gap can be varied by selection of appropriate exonuclease digestion conditions or by constructing heteroduplexes with specific gaps. The position of the gap can be varied by the procedures noted earlier under SITE SELECTION. Fourth, the method is limited by the availability of suitable mutagens. Bisulfite is an ideal reagent for this purpose, whereas other chemical mutagens are less suited since they generate products that base-pair ambiguously, or they are not specific for single-stranded polynucleotides, or they modify more than one nucleotide base. Perhaps enzymatic rather than chemical modification of specific bases will prove useful.

It should be noted that specifically gapped molecules are also suitable for incorporating base analogs during repair of the gap with DNA polymerase (18). Since the resulting duplex would have the normal, wild-type base opposite the analog and the analog must have ambiguous base-pairing properties, we would not expect mutants to be generated with efficiency comparable to that of the local mutageneous procedure just described. Nonetheless, coupled to appropriate selection either by phenotype or restriction enzyme resistance, it should be feasible to isolate a variety of mutants by base analog incorporation. Moreover, since different analogs can be used, this procedure is not limited to a single type of base-pair transition.

CLONING OF CONSTRUCTED MUTANTS

In terms of biological properties, SV40 mutants can be divided into those that are viable, i.e., able to grow and form a plaque, and those that are nonviable or helper-dependent, i.e., unable to form a plaque (without a complementing virus) under any attainable conditions. Viable mutants, whether deletion or point mutants, may be indistinguishable in biological properties from wild-type virus. They may be partially defective and thus form small or otherwise abnormal plaques or they may be conditionally defective (e.g., failing to form plaques at high or low temperature). Cloning of viable mutants is straightforward: cell monolayers are transfected with mutant DNA, and individual plaques are selected for analysis. If partially defective mutants are being sought, small or otherwise abnormal plaques are selected. For the isolation of temperature-dependent mutants, plaques picked at low temperature are tested at high temperature and vice versa. In all cases, it is essential to reclone each isolate by plating an appropriate dilution of the initial plaque suspension.

To isolate helper-dependent mutant viruses, the in vitro modified DNA is plated in the presence of a complementing helper virus by a procedure called complementation plaquing (19,20; Figure 6). If the helper virus is itself a defective mutant (e.g., a temperature-sensitive or deletion mutant), plaque formation will be dependent on coinfection of the same cell with the helper and the

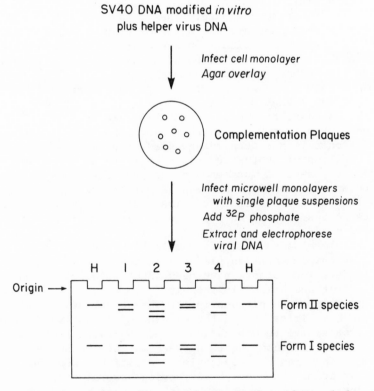

Figure 6.   Scheme for cloning and identifying helper-
            dependent mutants of SV40.

modified DNA and subsequent propagation of the complementing pair.
Obviously a helper must be selected that can complement mutants
defective in the gene of interest. Furthermore, since separation
of mutant and helper virus by buoyant density (21) or their DNAs
by electrophoresis (19) will likely be required for characterization
of the mutant, it is advantageous to use a helper whose genome
differs in length from that of the mutant, or one that can be
cleaved by a restriction enzyme to which the mutant DNA is resistant
(22). For example, to isolate small, constructed deletion mutants
or point mutants, a complementing helper with a large deletion
can be used, whereas to isolate large deletion mutants a small
deletion mutant or a point mutant (e.g., a ts mutant) can be used.
The resulting plaques are then surveyed for the presence of both
mutant and helper virus as described below.

SURVEY OF PLAQUES FOR MUTANTS

     To survey plaques for the presence of mutant viruses and for
initial characterization of their DNA, cells in microwells are infected with a small amount of plaque suspension, and viral DNA is
labeled with ^{32}P-orthophosphate. The extracted viral DNA is then
analyzed by electrophoresis and autoradiography, and by appropriate
restriction enzyme digestion. In the case of viable deletion
mutants, a short genome may be found by agarose electrophoresis
and/or an altered restriction pattern of its DNA may be seen. With
nonviable deletion mutants growing in the presence of helper virus,
both helper and mutant DNAs are found in the electrophoresis gel.
In the latter case, recovered mutant DNA does not yield plaques in
the absence of helper DNA. In the case of point mutants in which
a restriction site has been mutagenized, the mutant DNA is detectable by the loss of that site. In some instances, new sites are
created as a result of mutagenesis, as predicted from the nucleotide sequence surrounding the site of mutagenesis (15). For point
mutants that do not result in the loss or creation of a restriction site we must depend on a change in plaque morphology or temperature dependence of plaques to signal the presence of a mutant.
It is possible also to identify completely defective point mutants
isolated by complementation plaquing in the same way that small
nonviable deletion mutants have been identified, i.e., the presence
of both mutant and helper in the virus stock, and the failure of
mutant DNA to produce plaques in the absence of helper DNA.
     A serious limitation in the cloning of defective SV40 mutants
results from active recombination found in SV40 infected cells.
The appearance of recombinants between mutant and helper virus is
a frequence occurrence, detectable as additional DNA species in the
electrophoregram. However, by surveying several complementation
plaques during recloning of a mutant, and then making several stocks
from a given isolate, mutant stocks free of detectable recombinants

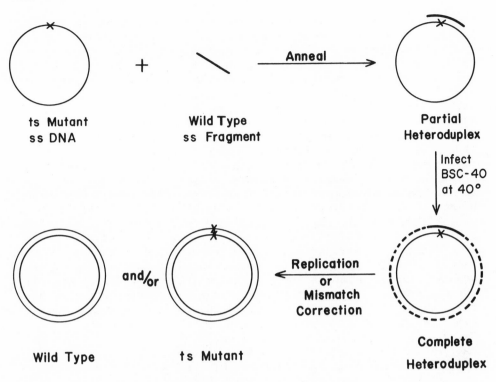

Figure 7.   Scheme for mapping SV40 mutants by marker rescue with
            fragments of viral DNA (25).

can generally be prepared. In a few instances, however, we have
been unable to prepare recombinant-free stocks.

MAPPING CONSTRUCTED MUTANTS

       The identification of deletion mutants and certain point mutants
during the survey of isolated clones just outlined often serves to
localize the mutational site with respect to the SV40 restriction
endonuclease cleavage map. In other instances, detailed restriction mapping of deletion mutant DNA recovered from agarose gels
will localize the mutational site, frequently within 1 to 2% of
the SV40 genome length. An additional useful method for deletion
mutants involves mapping heteroduplexes formed between full length
linear mutant and wild-type DNAs constructed by annealing restricted
form I DNA of each type. The resulting heteroduplexes are either
analyzed by electron microscopy (suitable for deletions greater
than about 100 nucleotides (23)), or by S1 nuclease digestion to

remove the deletion loop and cut the mutant DNA around the deletion joint, followed by electrophoretic sizing of the products (24).

In the case of partially or completely defective point mutants that do not involve the loss or gain of restriction sites, the mutation can be mapped by marker rescue with restriction fragments (25,26) as diagrammed in Figure 7. In this procedure, partial heteroduplexes between circular mutant DNA and a wild-type restriction fragment are used to infect cells under conditions where wild-type and mutant plaques are distinguishable. Only DNA fragments that correspond to the mutant site result in the formation of wild-type plaques, thus localizing the mutation to the map coordinates of the active fragment. The same procedure has also been applied to map small deletion mutants (27). It should be noted that mapping by marker rescue, as opposed to strictly physical mapping procedures, establishes that the mutation localized by the procedure is responsible for the biological defect.

Ultimately, mutational sites must be mapped at the nucleotide sequence level. Given present sequencing methods (28,29) and a detailed cleavage map of SV40, it is entirely feasible to determine the nucleotide sequence at the mutational site once that site has been localized to a region of about 200 base pairs or less by restriction mapping or marker rescue.

SEPARATION OF DEFECTIVE MUTANTS FROM HELPER VIRUS

Separation of defective SV40 mutant DNA from that of the helper virus is readily accomplished by agarose electrophoresis of the mixed DNAs provided there is a difference of about 5% or more in the length of the two DNA species (19). As a rule, prior conversion of DNA I to singly nicked DNA II leads to better purification since form I contains multiple electrophoretic bands. Moreover, depending on the particular mutant and helper, prior to electrophoresis it may be possible to selectively cleave the helper DNA with a restriction enzyme to which the mutant DNA is resistant (22).

The procedure for separating defective mutant and helper virions by buoyant density difference is less satisfactory, due primarily to low recovery of infectivity during the purification procedure. Nonetheless, where mutant and helper DNA differ by more than about 10% in length, three sequential bandings in CsCl (ρ = 1.34) provide mutant virus with about 0.1 to 1% contamination with helper virus (21). Addition of neutral detergent (e.g., 0.05% Triton X-100) to virus lysates and CsCl solutions results in more compact virion bands (Pipas, unpublished results). To quantitate infectious units of the mutant, a complementation assay is carried out with excess helper virus, yielding a measure of complementing units (21). Contaminating helper virus is similarly assayed with an excess of a complementing helper that does not complement the mutant.

SAMPLE PROTOCOLS

Site Selection

Single strand scission at the BglI site of SV40 DNA. Recently
prepared SV40 DNA form I at a concentration of 25 to 100 µg/ml was
incubated at 22°C with (final concentrations) 20 mM Tris-Cl, pH 7.6,
7 mM $MgCl_2$, 7 mM β-mercaptoethanol, 0.01% autoclaved gelatin, 80 µg/
ml ethidium bromide, and varying concentrations of BglI in a total
volume of 5 to 10 µl. The conversion of form I to form II DNA was
monitored by stopping the reaction with 1/10 volume of 100 mM EDTA/
3% sodium dodecylsulfate/50% sucrose/0.02% bromphenol blue, and
electrophoresing in a 1.4% agarose gel slab. The reaction was then
scaled up using a concentration of BglI that converted about 90%
of form I DNA to form II. The preparative reaction mixture was
made 10 mM in EDTA, extracted once with phenol saturated with 0.1 M
Tris-Cl, pH 8.0, and dialyzed against 2 mM Tris-Cl, pH 8.0-0.2 mM
EDTA. This DNA was used directly for subsequent gapping by
exonuclease.

Nick translation from the BglI site with E. coli DNA polymerase
I (31). Controlled nick translation can be carried out with low
concentrations of all four deoxynucleoside triphosphates, or (to
stop at a specific nucleotide), with three of the four deoxynucleoside triphosphates (Figure 3). A typical preparative reaction (100
to 200 µl) contained 20 µg/ml BglI-nicked DNA, 50 mM potassium
phosphate, pH 7.4, 10 mM $MgCl_2$, 0.5 µM of each deoxynucleoside triphosphate (including tracer amount of one α-^{32}P-dNTP) and 30 units
of endonuclease-free E. coli DNA polymerase I (Worthington Biochemical Corp., Freehold, NJ). After incubation at 14°C for 5 to 10
min, the reaction was stopped by addition of EDTA to 20 mM followed
by phenol extraction. Samples were assayed for cold acid precipitable
radioactivity (32) and (by BglI cleavage and acrylamide gel electrophoresis under denaturing conditions (33)) for the extent and syn-
chrony of nick translation. Under these conditions, approximately
10 nucleotides per min were added to the 3' end of DNA at the original BglI nick. Judging by the breadth of the electrophoretic bands
of ^{32}P-single-stranded restriction fragments, the reaction remained
well synchronized (± about 20 nucleotides) in both directions for
at least 10 min. Prior to proceeding with the further enzymatic
treatment of the nick-translated DNA, the phenol-extracted material
was passed through Sephadex G-50 and dialyzed against 1.5 mM NaCl-
0.15 sodium citrate, pH 7.0. It should be noted that since the
original BglI-nicked DNA consists of two populations of molecules --
those nicked on the E strand and those on the L strand -- nick
translation also occurs on either the E or L strand.

Single cleavage with HindIII, a multicut enzyme for SV40 (5).
Each batch of enzyme must be titrated under given conditions of DNA I
and ethidium bromide concentrations, and temperature and time of
incubation to optimize the yield of full length DNA ($L_{HindIII}$).

Maximal yield of $L_{HindIII}$ was obtained when about half of the product was form II DNA and half was full length linear DNA as determined by agarose gel electrophoresis. Form I DNA (100 to 200 µg/ml)
was incubated in a solution containing 6 mM Tris-Cl, pH 7.5, 100 mM
NaCl, 6 mM $MgCl_2$, 6 mM β-mercaptoethanol, 0.01% autoclaved gelatin,
10 µg/ml ethidium bromide and sufficient HindIII to give an enzyme
to DNA ratio of 1.5 units (Bethesda Research Laboratories, Rockville,
MD) per µg of DNA. After incubation at 37° for 30 min, EDTA addition, and phenol extraction of the DNA, form II and linear DNA were
purified by agarose gel electrophoresis.

Cyclization of Linear DNA

     In vitro cyclization is done by blunt-end ligation of linear
DNA (12). Generally, blunt-ends must first be created by removing
3' or 5' overhangs at the ends of linear DNA. For this purpose,
the single-strand-specific nuclease S1 has been used (34). We have
used a concentration of S1 sufficient to digest completely an
equivalent amount of denatured SV40 DNA, having determined that
under the conditions used, duplex linear SV40 DNA is not digested.
In a typical reaction, 300 to 400 µl volume, 10 to 20 µg/ml of
linear DNA was incubated with 30 mM sodium acetate, pH 4.4, 270
mM NaCl, 0.5 mM $ZnSO_4$ and 2.5 units of S1 per µg of DNA at 25°C
for 15 min. After addition of Tris-Cl, pH 8.6, to 50 mM and phenol
extraction, the DNA was dialyzed against 1.5 mM NaCl-0.15 mM sodium
citrate. For ligation, T4 DNA ligase (11) (Miles Research Products,
Elkhart, IN or P-L Biochemicals, Milwaukee, WI) was used after ti-
tration of each batch of enzyme for activity in circularizing the
S1-treated linear DNA (electron microscopy assay). In a typical
reaction of 10 to 50 µl, 6 to 8 µg/ml of S1-treated DNA was incubated
with 50 mM Tris-Cl, pH 8.0, 10 mM $MgCl_2$, 0.1 mM EDTA, 13 mM dithiothreitol, 1.0 mM ATP and 1 to 5 units of T4 DNA ligase at 15°C for
16 to 20 hr. The extent of cyclization was assessed by microscopy
before terminating the reaction. Under these conditions, about 50%
of the DNA molecules were cyclized. If removal of residual linear
DNA is desired, this can be done with least manipulative loss by
treatment with the ATP-dependent DNase of H. influenzae (13),
which requires free duplex ends for its degradative activity. For
this purpose, the ligation reaction was terminated by heating at
60°C for 5 min and subsequently incubated with the above DNase
(20 units/ml) at 37°C for 20 min. Electron microscopic monitoring
showed that only circular molecules remained. After addition of
EDTA, phenol extraction and dialysis, the DNA was used for
transfection.
     For cell-mediated cyclization of linear DNA, the protocol for
cloning of mutants from in vitro-modified DNA has been used (see
below). Where extended deletions are desired, complementary (cohesive) single-strand tails can be removed with S1 nuclease prior
to transfection as described above.

Local Mutagenesis

Gapping with M. luteus DNA polymerase (15). To generate a short, single-stranded gap extending predominantly in the 5' to 3' direction from a nick, the exonuclease activity of DNA polymerase I from Micrococcus luteus (Miles) was used. Twenty to 100 µg/ml of singly-nicked form II SV40 DNA in a volume of 50 to 100 µl was incubated with 70 mM Tris-Cl, pH 8.0, 7 mM $MgCl_2$, 2 mM β-mercaptoethanol, 0.6 mM dTTP and 1 unit of M. luteus DNA polymerase I (35) per µg DNA at 11°C for 1 hr. Under these conditions and starting with HpaII-nicked SV40 DNA, the 5' to 3' exonuclease activity of DNA polymerase appeared to release 5 or 6 nucleotides per molecule. If the reaction proceeds from a nick at a one-cut restriction enzyme site, the formation of a gap eliminates the site and can be followed by the appearance of form II resistant to cleavage by the restriction enzyme. Alternatively, changes in the S1 nuclease sensitivity of the form II DNA can be monitored; the fraction of form II DNA linearized by S1 corresponds to the fraction possessing a single-stranded gap. For subsequent use of the gapped DNA, the reaction mixture was made 10 mM in EDTA, heated for 10 min at 65°C, phenol-extracted and dialyzed against 15 mM NaCl-1.5 mM sodium citrate, pH 7.0.

Gapping with exonuclease III of E. coli. Short gaps of varying length in the 3' to 5' direction can be efficiently generated with exonuclease III from E. coli (36). One hundred µg/ml of singly-nicked SV40 DNA II in a volume of 50 to 100 µl was incubated at 14°C with 50 mM Tris-Cl, 5 mM $MgCl_2$, 5 mM dithiothreitol, 0.01% bovine serum albumin and variable amounts of exonuclease III. Since the reaction of exonuclease III duplex DNA is linear with time, the number of nucleotides excised can be controlled by establishing a kinetic curve of digestion at several enzyme concentrations and then extrapolating to the appropriate time. We selected conditions under which approximately 1% of the DNA (equivalent to about 100 nucleotides) was released as acid soluble counts from HpaII-nicked SV40 DNA per 30 min at 14°C. Gel electrophoresis under denaturing conditions of restriction fragments containing the HpaII site demonstrated that digestion was approximately synchronous for at least 10 min, i.e., molecules with gaps of about 20 nucleotides or less could be prepared readily. Prior to subsequent mutagenesis the gapped DNA was treated as described above.

Reaction of gapped DNA with sodium bisulfite (15). All reactions of gapped DNA with sodium bisulfite were carried out under a standard set of conditions; to control the extent of reaction, only the time of incubation was varied. The reaction mixture was as follows: 3 volumes of 4 M sodium bisulfite, pH 6.0 (312 mg $NaHSO_3$, 128 mg Na_2SO_3, 0.86 ml deionized H_2O), 0.04 volumes of 50 mM hydroquinone, 1 volume of form II DNA (10 to 100 µg/ml) in 15 mM NaCl-1.5 mM sodium citrate, pH 7.0 and finally paraffin oil overlay. All reagents were prepared immediately prior to use and cooled on ice before mixing in the order listed above. The mixture was

incubated at 37°C in the dark. To terminate the reaction, the incubation mixture was dialyzed against the following sequence of
buffers: 1) 1000 volumes of 5 mM potassium phosphate, pH 6.8/0.5 mM
hydroquinone at 0°C for 2 hr; 2) repeat of 1); 3) 1000 volumes of
5 mM potassium phosphate, pH 6.8, at 0°C for 4 hr; 4) 1000 volumes
of 0.2 M Tris-HCl, pH 9.2/50mM NaCl/2 mM EDTA at 37°C for 16 to 24
hr; 5) 1000 volumes of 2 mM Tris-HCl, pH 8.0/2 mM NaCl/0.2 mM EDTA
at 4°C for 6 to 12 hr. The water used to prepare the first four
dialysis buffers was degassed by vigorous boiling prior to use.
Under the above conditions, about 8% of C residues in single-
stranded DNA were deaminated after 1 hr of incubation, and about
30% after 4 hr.

     At this stage either the mutagenized form II DNA was used
directly to transfect BSC-40 cells by the DEAE-dextran method (37),
or the single-stranded gap was first repaired with DNA polymerase
I plus the four deoxynucleoside triphosphates before transfection
(15). If the objective is to construct point mutations that
eliminate a one-cut restriction enzyme site, repair of the gap
permits an in vitro selection for such mutants; the repaired form
II is subjected to a final digestion with the one-cut enzyme, and
the resistant form II, after separation from other DNA species by
agarose gel electrophoresis, is used for transfection.

                    Cloning and Identification of Mutants

     Mutants with partial or conditional defects. In vitro-modified
SV40 DNA, i.e., linearized or locally mutagenized DNA, can be used
directly to infect monolayers of permissive monkey cells. By
infecting at 32°, 37° and 40° and replating individual plaques at
all three temperatures, mutants with temperature-dependent growth
and those that are partially defective can be detected. As an
example, we describe the cloning and identification of BglI-resistant
mutants from DNA mutagenized with bisulfite at the BglI site (15).
Six cm dishes of BSC-40 cells were infected with from 0.2 to 2.0 mg
of modified DNA using the DEAE dextran method, and the monolayer
was overlayed with agar, as in the SV40 plaque assay. At 37°,
approximately 50 plaques per ng of DNA were detected, including
wild-type, small and occasional sharp-edged plaques. All plaques
selected at 37° were replated at 32°, 37° and 40°. Four plaque
morphology phenotypes were observed: 1) wild-type at all three
temperatures, 2) small, sharp plaques at all temperatures, 3) tiny
plaques at 32° and 40° and small plaques at 37°, and 4) small
plaques at 32° but wild-type at 37° and 40°. To test each mutant
type for loss of the BglI site, 16 mm microwell monolayers of
BSC-40 cells were infected with a drop of plaque suspension and
^{32}P-viral DNA prepared (see below). Viral DNA labeled with ^{32}P
was then incubated with BglI in the presence of excess, nonlabeled
wild-type (BglI-sensitive) DNA, and electrophoresed in 1.4% agarose
gels. BglI sensitivity of ^{32}P DNA was assessed by autoradiography,

and the activity of the enzyme by ethidium bromide staining of the gel. Mutants of all four plaque morphology classes were found to have BglI-resistant DNA.

Helper-dependent mutants (19,20). The procedure for cloning and identifying helper-dependent mutants of SV40 from suitably modified DNA is diagrammed in Figure 6. As an example of its application, the isolation of early deletion mutants of SV40 from S1-treated, full length linear molecules produced by HindIII (L$_{HindIII}$ (S1)) in the presence of ethidium bromide is described. The complementing helper in this case was a SV40 late mutant (dl-1007, Ref. 3) with a large deletion (24% of the genome) that takes out part of the VP 1, 2 and 3 genes. Thus, the helper can complement only early mutants and can be separated from such mutants if their deletions are less than about 15% of genome length. DNA of dl-1007 was prepared by two consecutive electrophoretic separations from the DNA of its own helper, a tsA mutant of SV40 (3). It was then tested for biological purity by infecting BSC-40 cell monolayers in the presence or absence of tsA DNA at 40°. In the absence of tsA DNA, 30 ng yielded no plaques (at 37° or 40°) whereas in the presence of 50 ng of tsA DNA, >500 plaques were observed.

To isolate HindIII early deletion mutants, subconfluent BSC-40 monolayers in 6 cm dishes were infected at 37° with a mixture of L$_{HindIII}$ (S1) (0.2 to 1 ng) and dl-1007 DNA (30 ng) by the DEAE dextran method. After allowing plaques to develop under agar in the usual way, individual plaques were picked into 0.5 ml of medium. To screen plaques for mutants, BSC-40 cells in 6 mm microwells (96 wells per plate) were infected with 50 μl of plaque suspension, and at 96 hr post infection the medium was replaced by low phosphate medium containing 5 μCi of ^{32}P-orthophosphate. One day later, an additional 5 μCi of ^{32}P-orthophosphate was added, and after another day, the cells were lysed and viral DNA extracted by the method of Hirt (38), using 100 μl of lysing solution per well. After RNase treatment, phenol extraction and alcohol precipitation (in the presence of 10 μg of tRNA), the precipitated DNA was taken to dryness and dissolved in 25 μl of 15 mM NaCl-1.5 mM sodium citrate. About 5 to 10 μl of each sample was then electrophoresed in a 1.4% agarose slab gel with 40 to 50 slots, and the slab was autoradiographed wet (where DNA was to be recovered) or after drying. Plaques containing cloned defective deletion mutants will each yield a DNA species corresponding to that of dl-1007 plus one other DNA species. Generally the dl-1007 DNA (containing the intact early gene) is at least as abundant as the early deletion mutant DNA. To localize the early deletion, mutant DNA recovered from the gel was digested with HindIII and analyzed by electrophoresis in acrylamide gel. Mutants of interest were then selected for processing.

Before preparation of a mutant stock, the original plaque suspension or a microwell stock derived from it was replaqued at least once. Again, individual plaques were screened as described above to ensure that a replaqued mutant yielded only mutant and dl-1007 DNA. (Since recombinants arise with substantial frequency, it is necessary to screen every lysate.) To prepare primary stocks, 50 μl of screened replaqued virus suspension was used to infect the edge of a 6 cm dish of subconfluent BSC-40 cells, and after absorption for 1 hr at 37°, fresh medium was added. Lysis generally was complete in 10 to 16 days at 37°. Each primary stock was screened as described earlier, except that infected cells were labeled at 24 and 48 hr post infection, and viral DNA was extracted at 72 hr. Stocks, free of recombinants, were then used to prepare secondary stocks, generally by infecting a 75 cm^2 flask of BSC-40 cells with 1 ml of a 1:5 dilution of primary stock.

Mapping SV40 Mutants by Marker Rescue (25)

A diagram of the method is shown in Figure 7. First, mutant form I DNA was converted to singly-nicked form II. A convenient procedure is to use DNase I in the presence of ethidium bromide: 2 to 10 μg of mutant form I DNA in a final volume of 50 to 100 μl was incubated with 50 mM Tris-Cl, pH 7.2, 5 mM MgCl$_2$, 150 μg/ml ethidium bromide and 25 to 100 ng/ml pancreatic DNase at room temperature for 30 min. (As described in the section on specific nicking with the BglI restriction enzyme, the exact amount of nuclease sufficient to convert 90 to 100% of form I to form II must be determined in a series of pilot reactions.) Second, appropriate restriction fragments from wild-type SV40 DNA were prepared. As noted elsewhere (25), it is advisable to purify fragments by two sequential enzyme digestions. Next, heteroduplexes between denatured mutant form II DNA and a wild-type restriction fragment were formed in a denaturation-partial reannealling mixture made up of the following: 50 ng singly-nicked mutant form II DNA and a 15- to 20-fold molar excess of a wild-type restriction fragment in 150 μl, 20 μl of 1 N NaOH, and after 10 min at room temperature, 30 μl of 0.67 N HCl/0.33 M Hepes, pH 7.4. After incubation of this mixture at 65°C for 15 min and cooling to room temperature, the DNA was used in the standard transfection assay on BSC-40 cells. Following incubation of infected dishes under conditions that distinguish mutant and wild-type plaques, the appearance of wild-type plaques indicates that the mutation responsible for the plaquing defect maps within the genomic region corresponding to the active restriction fragment. Tests with other fragments serve as controls.

Acknowledgments: The authors' research, described in this review, has been supported by grants from the National Cancer Institute, U.S. Public Health Service (CA16519), the American Cancer Society and the Whitehall Foundation.

REFERENCES

1   Reddy, V.B., Thimmappaya, B., Dhar, R., Subramanian, K.N.,
    Zain, B.S., Pan, J., Ghosh, P.K. and Weissman, S.M. (1978)
    Science 200, 494–502.
2   Fiers, W., Contreras, R., Haegeman, G., Rogiers, R., Van de
    Voorde, A., Van Heuverswyn, H., Van Herreweghe, J., Volckaert,
    G. and Ysebaert, M. (1978) Nature 273, 113–120.
3   Lai, C.-J. and Nathans, D. (1974) J. Mol. Biol. 89, 179–193.
4   Carbon, J., Shenk, T.E. and Berg, P. (1975) Proc. Nat. Acad.
    Sci. U.S.A. 72, 1392–1396.
5   Parker, R.C., Watson, R.M. and Vinograd, J. (1977) Proc. Nat.
    Acad. Sci. U.S.A. 74, 851–855.
6   Mertz, J.E., Carbon, J., Herzberg, M., Davis, R.W. and Berg,
    P. (1975) Cold Spring Harbor Symp. Quant. Biol. 39, 69–84.
7   Lai, C.-J. and Nathans, D. (1976) Virology 75, 335–345.
8   Shenk, T.E., Carbon, J. and Berg, P. (1976) J. Virol. 18,
    664–671.
9   Nathans, D., Lai, C.-J., Scott, W.A., Brockman, W.W. and
    Adler, S.P. (1977) in Genetic Manipulation as it Affects the
    Cancer Problem (Scott, W.A., ed.), pp. 1–10, Academic Press,
    New York, NY.
10 Dugaiczyk, A., Boyer, H.W. and Goodman, H.M. (1975) J. Mol.
    Biol. 96, 171–184.
11 Weiss, B., Jacquemin-Sablon, A., Live, T.R., Fareed, G.C. and
    Richardson, C.C. (1968) J. Biol. Chem. 243, 4543–4555.
12 Sgaramella, V., van de Sande, J.H. and Khorana, H.G. (1970)
    Proc. Nat. Acad. Sci. U.S.A. 67, 1468–1475.
13 Wilcox, K.W. and Smith, H.O. (1976) J. Biol. Chem. 251, 6127–
    6134.
14 Mertz, J. and Davis, R.W. (1972) Proc. Nat. Acad. Sci. U.S.A.
    69, 3370–3374.
15 Shortle, D. and Nathans, D. (1978) Proc. Nat. Acad. Sci.
    U.S.A. 75, 2170–2174.
16 Shapiro, R., Braverman, B., Louis, J.B. and Servis, R.E.
    (1973) J. Biol. Chem. 248, 4060–4064.
17 Hayatsu, H. (1976) Progr. Nucl. Acid Res. Mol. Biol. 16,
    75–124.
18 Müller, W., Weber, H., Meyer, F. and Weissmann, C. (1978)
    J. Mol. Biol. 124, 343–358.
19 Brockman, W.W. and Nathans, D. (1974) Proc. Nat. Acad. Sci.
    U.S.A. 71, 942–946.
20 Mertz, J.E. and Berg, P. (1974) Virology 62, 112–124.
21 Scott, W.A., Brockman, W.W. and Nathans, D. (1976) Virology
    75, 319–334.
22 Cole, C.N., Landers, T., Goff, S.P., Manteuil-Brutlag, S.
    and Berg, P. (1977) J. Virol. 24, 277–294.
23 Davis, R.W., Simon, M. and Davidson, M. (1971) Methods
    Enzymol. 21, 413–427.

24 Shenk, T.E., Rhodes, C., Rigby, P.W.J. and Berg, P. (1975)
    Proc. Nat. Acad. Sci. U.S.A. 72, 989–993.
25 Lai, C.-J. and Nathans, D. (1975) Virology 66, 70–81.
26 Mantei, N., Boyer, H.W. and Goodman, H.M. (1975) J. Virol.
    16, 754–757.
27 Feunteun, J., Sompayrac, L., Fluck, M. and Benjamin, T.
    (1976) Proc. Nat. Acad. Sci. U.S.A. 73, 4169–4173.
28 Maxam, A.M. and Gilbert, W. (1977) Proc. Nat. Acad. Sci.
    U.S.A. 74, 560–564.
29 Sanger, F., Nicklen, S. and Coulson, A.R. (1977) Proc. Nat.
    Acad. Sci. U.S.A. 74, 5463–5467.
30 Kelly, T.J. Jr. and Nathans, D. (1977) Adv. Virus Res.
    21, 85–173.
31 Kelly, R.B., Cozzarelli, N.R., Deutscher, M.P., Lehman,
    I.R. and Kornberg, A. (1970) J. Biol. Chem. 245, 39–45.
32 Englund, P.T. (1972) J. Mol. Biol. 66, 209–224.
33 McMaster, G.K. and Carmichael, G.G. (1977) Proc. Nat. Acad.
    Sci. U.S.A. 74, 4835–4838.
34 Vogt, V.M. (1973) Eur. J. Biochem. 33, 192–200.
35 Miller, L.K. and Wells, R.D. (1972) J. Biol. Chem. 247,
    2667–2674.
36 Richardson, C.C. (1966) in Procedures in Nucleic Acid Re-
    search (Canton, G.L. and Davies, G.R., eds.), pp. 212–223,
    Harper & Row, New York, NY.
37 McCutchan, J.H. and Pagano, J.S. (1968) J. Nat. Cancer Inst.
    41, 351–356.
38 Hirt, B. (1967) J. Mol. Biol. 26, 365–369.

STRUCTURE OF CLONED GENES FROM XENOPUS: A REVIEW

R.H. Reeder[*]

Department of Embryology
Carnegie Institution of Washington
115 West University Parkway, Baltimore, Maryland 21210

Xenopus

This chapter will summarize our current knowledge concerning
the structure and function of the four gene families that have so
far been cloned from the frog genus, Xenopus. Three of these gene
families are present in hundreds to thousands of copies in the
genome and code for stable structural RNAs. These are the genes
coding for the large 7.5 kb ribosomal RNA precursor molecule
(termed ribosomal DNA or rDNA), the genes coding for various 5S
ribosomal RNAs (termed 5S DNA) and the genes coding for the initiator methionyl-tRNA (termed $tDNA_1^{met}$). The fourth gene family we
will consider is the small set that codes for vitellogenin, the
large precursor to yolk protein.

A previous review on rDNA and 5S DNA has covered the field up
to the introduction of molecular cloning and restriction enzyme
technology (1). This article will concentrate on information
accumulated since that time.

RIBOSOMAL DNA

The rDNA of X. laevis occupies a unique niche in the history
of eukaryotic molecular biology. It was the first eukaryotic gene
to be isolated and to have its structure examined in detail (2,3).
It was also the first to be recombined in vitro with a bacterial
plasmid and cloned (4). Therefore, it seems appropriate to commence
this article by considering the structure of rDNA.

---

* Present address, Hutchinson Cancer Research Center, 1124 Columbia
  Street, Seattle, Washington 98104.

X. laevis rDNA is an example of a multigene family. Approximately 450 tandemly duplicated copies of the gene are situated at a single locus represented once per haploid set of chromosomes. The repeating unit of rDNA consists of a nontranscribed spacer region that can vary in length from 2.7 kb to over 7 kb plus a transcribed gene region of 7.5 kb which is transcribed as a single molecule. The primary transcript is then processed (5) to yield mature rRNA molecules of 18S (2 kb), 5.8S (161 bases) and 28S (4 kb). Transcription of the precursor proceeds in the order 5' 18S → 5.8S → 28S 3' (6-8). A diagram of one repeating unit of rDNA is shown in Figure 1.

Heterogeneity of the Nontranscribed Spacers

The repeating units of rDNA are closely related to each other but they are not all identical. Conclusive evidence that the repeating units have variable lengths was first provided by Morrow et al. (4). To construct the first recombinant clones, rDNA was digested with the restriction enzyme EcoRI and from the unexpected complexity of the fragments, Morrow and coworkers correctly deduced that the repeating units were variable in length. They further proposed that the variability resided most likely in the nontranscribed spacer. At about the same time, Wellauer et al. (9) used a combination of EcoRI digestion, heteroduplex formation, and secondary structure mapping on uncloned genomic rDNA to conclusively localize the heterogeneity in the nontranscribed spacer.

Our present knowledge of rDNA spacer structure and biology is largely based on studies of four cloned spacer-containing fragments which were generated by EcoRI digestion (Figure 1). The spacers in these clones range in length from the shortest normally observed (2.7 kb) to one of the largest (7 kb).

A major portion of each spacer is composed of simple sequence repetitive DNA and long spacers which differ from shorter spacers by having more of the repetitive elements. This arrangement was initially deduced by electron microscopy of heteroduplexes between long and short spacers (10). The excess DNA in the longer spacers invariably appeared as deletion loops. Because of the repetitive nature of the DNA, the location of deletion loops was not fixed but was observed to vary within boundaries. Determination of these boundaries led to the conclusion that each spacer could be subdivided into at least three regions (Figure 1). Region A was next to the 3' end of the gene, was about 500 bases long, and was not internally repetitive. The rest of the spacer was divided into two domains, B and D, each of which was internally repetitive. Originally, the boundary between B and D was also given its own designation as region C. Mapping of restriction enzyme sites within the same four spacers (11) confirmed that the electron microscopic deductions were generally correct. However, the boundary between

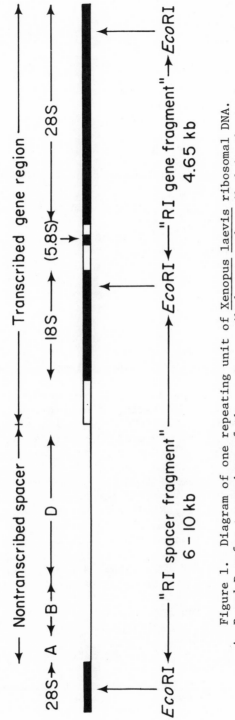
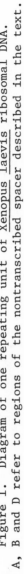

Figure 1. Diagram of one repeating unit of Xenopus laevis ribosomal DNA.
A, B and D refer to regions of the nontranscribed spacer described in the text.

regions B and D does not appear to be a discrete stretch of DNA
but simply a boundary where one type of repetitive sequence stops
and a different one begins. Therefore, the region C designation
has been discarded.

Both heteroduplex mapping (10) and restriction mapping (11)
agree that regions B and D are distinguishable from each other.
Cross-hybridization experiments demonstrate, however, that the
regions share a detectable level of sequence homology. It is
possible, therefore, that at one time an ancestral repetitive
sequence stretched across the entire spacer. Some as yet unknown
event then formed a boundary in the middle of the spacer allowing
the two halves to evolve apart.

Heteroduplexes between rDNA from X. laevis and X. borealis
(R. Reeder, unpublished data) show that the borealis spacer has
the same bipartite structure. (In earlier publications on rDNA
and 5S DNA, Xenopus borealis was mistakenly identified as Xenopus
mulleri (12)). Region D still has residual homology between the
two species. Regions A and B, however, have diverged such that
no homology is detectable.

In the four cloned spacers that have been studied in detail,
region B remains fairly constant in length while region D accounts
for most of the overall length variation. Heteroduplex formation
between a cloned spacer and uncloned amplified rDNA shows, however,
that region B is also capable of expansion (13).

Repeating units with spacers of different lengths can be intermingled on the same chromosome although there is some controversy
concerning the degree of clustering of identical length spacers.
Wellauer et al. (10) formed heteroduplexes between a single type
of cloned spacer and long stretches of uncloned rDNA. By examining the frequency with which the nearest neighbor spacers of unequal length occurred, they concluded that in somatic, chromosomal
rDNA there was extensive, perhaps random, intermingling of different
spacer classes. Because of the laboriousness of the assay, however, they only examined rDNA from two different frogs. Buongiorno-
Nardelli et al. (14) have recently screened a number of animals by
partially digesting their rDNA with EcoRI and then comparing the
distribution of fragment sizes with the patterns expected whether
clustering was present or absent. They concluded that in chromosomal rDNA there was extensive clustering. It could be argued,
however, that an EcoRI partial digest is complex and the crucial
fragments are of a size that is not well resolved on agarose gels.
A better enzyme to use is HindIII, which cuts only once per repeating unit. Kirkman and Southern (15) have used this enzyme to
screen a number of animals and report no apparent clustering of
same-size spacers.

It seems clear from the foregoing discussion that various
spacer sizes can be highly intermingled on a single chromosome.
It would not be too surprising, however, if the degree of interspersion varied from frog to frog. The question is of some interest

because of what it may tell us about the mechanism of rDNA amplification in <u>Xenopus</u> oocytes. During amplification, extrachromosomal copies of rDNA are produced. Using the same nearest neighbor analysis mentioned above, Wellauer et al. (13) concluded that there was nearly complete clustering of spacer size classes in the amplified rDNA as opposed to nearly random interspersion in the chromosomal rDNA. This led them to propose that the initial amplification event produced an extrachromosomal copy of a single repeating unit. By way of rolling circle intermediates, this single copy was then amplified to form a population of uniform, highly clustered repeats.

More recently, Bird (16) has analyzed the spacer heterogeneity present in amplified rDNA from single oocytes. The female studied had approximately eight different spacer lengths present in the chromosomal rDNA isolated from blood cells. In contrast, most individual oocytes contained primarily one spacer size (although often a different size from oocyte to oocyte) along with traces of other size classes. This observation argues strongly that the initial amplification event removes only one size class at a time from the chromosome. Otherwise at least some of the classes would be present in equal amounts in the same oocyte. If spacer lengths are randomly intermingled on the chromosome, the most likely situation is that a single repeat is amplified at a time. Bird's results show, however, that the primary event can happen more than once per oocyte, even though one repeat length often gets a head start (16).

The overall spacer distribution in the whole ovary of a single frog is the sum of spacer abundances in each oocyte. Therefore, it is perhaps not surprising that the overall spacer pattern in amplified rDNA often shows a strong bias toward one or a few spacer sizes. It is remarkable, however, that this bias is reproducible among siblings and can be inherited from frog to frog (17).

Biochemical studies have emphasized the strong clustering of identical spacer lengths in amplified rDNA. It should be mentioned, however, that electron microscope studies of transcriptionally active rDNA have reported seeing unequal nearest neighbors (18). It is difficult to know whether they arise from the primary amplification embracing more than one repeat or from later crossing over between extrachromosomal molecules.

Since the rDNA at one nucleolar organizer locus often has a spacer pattern different from that at other loci, the spacer pattern can be used as a genetic marker. Reeder et al. (17) have examined the spacer patterns in a total of 50 frogs resulting from three separate matings. In general, the spacer pattern was transmitted unchanged from parent to offspring. In two cases an altered spacer pattern was seen in the progeny but it was not possible to determine the cause of the change. Crossing over between homologs should have been detectable in these experiments while sister chromatid exchange probably would not have been detected.

Methylation of rDNA

The chromosomal rDNA in somatic cells has 13% of its cytosines methylated as 5-methylcytosine while amplified rDNA in oocytes is unmethylated (3). In vertebrate DNA, methylation occurs almost exclusively on the doublet CpG. It is also known that the recognition sequences for several restriction enzymes contain the CpG doublet and that methylation of the cytosine blocks the cutting activity of the enzyme. Bird and Southern (19) and Bird (20) have made use of these facts to learn something about the organization of methylation sites in chromosomal rDNA.

They used the enzymes HpaII, HhaI, AvaI and HaeII, all of which have a recognition sequence containing CpG. As expected, amplified rDNA was cut into a large number of fragments by these enzymes while chromosomal rDNA was almost completely resistant. They were able to calculate that most of the possible recognition sites in chromosomal rDNA are methylated to the extent of 99% per site. Two relatively sensitive sites were found, one in the 28S coding region and one in region D of the nontranscribed spacer. The possible significance of these sites is at present unknown.

The CpG doublet is self-complementary. Bird (20) was able to demonstrate that methylation is present on both strands to the same degree. By a combination of DNA density labeling and radio-active methyl labeling he also concluded that the pattern of methyl labeling is inherited upon DNA replication (20).

Transcription of Ribosomal DNA In Vivo

In living cells, rDNA is transcribed by RNA polymerase I, a polymerase that is resistant to high levels of the fungal toxin, α-amanitin (21). The earliest transcript of rDNA that can be detected in pulse-labeled cells is the so-called 40S precursor rRNA, a molecule of 7.5 kb (22). In addition, micrographs of tran-scriptionally active rDNA (18,23) generally show transcription complexes with a short to long polarity covering a region of about 7.5 kb (assuming the DNA is in the noncontracted B form). These two observations support the assumption that the 7.5 kb precursor is the primary transcript of Xenopus rDNA. A more rigorous proof would be to demonstrate the presence of a poly-phosphate terminus on the 5' end of the precursor. This has recently been accomplished by Reeder et al. (24) who showed that up to 20% of the 7.5 kb molecules isolated from oocyte nucleoli could be capped in vitro by capping enzymes from Vaccinia. These enzymes have been shown to absolutely require a di- or triphosphate terminus for activity (24,25). By hybridizing radioactively capped precursor to various rDNA restriction fragments, the entire repeating unit has been searched for possible transcription initi-ation sites. Only two potential sites have been located so far:

one at the 5' end of the 7.5 kb sequence and the other somewhere
within the 4.6 kb fragment that EcoRI excises from the middle of
the transcribed gene. It is not yet known whether the latter site
represents a vestigial initiation site or just a chance homology.

No initiation sites were detected anywhere in the nontranscribed
spacer region. How can we then account for the occasional presence
of transcription complexes in this region (18)? It is possible
that RNA polymerase does initiate in the nontranscribed spacer but
the transcripts are too rapidly degraded to be detected in the
capping assay. A more probable explanation is that termination
does not always function properly.

Whatever the explanation may be, the weight of the evidence
argues that transcription normally begins at the 5' end of the 7.5
kb sequence. Hybridization of radioactively capped precursor to
restriction fragments has localized the initiation site to a 216
bp SmaI fragment (Figure 2) (Sollner-Webb and Reeder, unpublished
data). Recovery of the labeled RNA and digestion with RNAse A
resulted in the recovery of two different capped 5' oligonucleotides
suggesting that the 7.5 kb molecules are heterogeneous at their 5'
ends. The nature of this heterogeneity has not yet been completely
established. A sequence of 360 bases surrounding the initiation
site has been sequenced from one rDNA clone (pXlr14) (Figure 2)
(Sollner-Webb and Reeder, unpublished data). When the SmaI 216-mer
from pXlr14 was hybridized with 7.5 kb rRNA, S1 nuclease digestion
yielded only a single length of protected DNA whose terminus cor-
responded to the first A (nucleotide +1 in Figure 2) of the sequence
AAG. Since the majority of 7.5 kb rRNA begins with the sequence
AAG, it is probable that this sequence marks the transcription initiation site for most 7.5 kb molecules. It will require further
sequencing of other rDNA clones (as well as sequencing of the 5'
end of the 7.5 kb precursor) to establish the nature and degree
of heterogeneity in the initiation sequence.

Immediately to the left of the putative initiation site shown
in Figure 2 is a region of about 40 nucleotides that, by analogy
with X. laevis 5S DNA (see below) and prokaryotic genes, has a high
probability of being the promoter region. At present three things
can be said about this region. 1) It has a lower GC content (50%)
than do regions immediately to the left (75%) and to the right (80%).
2) It has no significant dyad symmetries or direct repeats. 3) It
has no apparent homologies with any of the promoters that are
recognized by X. laevis RNA polymerase III (26).

To the left of the initiation site on pXlr14 is a region of
about 200 bases that is relatively well conserved in sequence within
the species laevis and has no obvious repetitive structure. Left-
ward beyond that is region D of the simple sequence spacer. Whether
this 200-base region has any regulatory function remains to be seen.
There is also a similar (i.e., conserved, nonrepetitive) sequence
of 500 bases just outside the gene on the 3' end (11).

```
      -220      -210      -200      -190      -180      -170      -160      -150      -140
▼CCGGGGCCGCCGGGGCCCCGGCCGAGCAAGTCCCCGCGCGGGGCGCCCACTTTGCGAAAAAACGGGCGCCAAAGGGCCCGGGCCCTCCCGCG▼

      -130      -120      -110      -100       -90       -80       -70       -60       -50
GAGGCCCCGATGAGGACGGATTCGCCCGCGCCCCGCGCCCGGAGTTCCGGAGCCCGGGGAGAGGAGCCGGCGCCGGCTCTCGGCCCCG
                                                       ▼

       -40       -30       -20       -10                  10        20        30        40
CACGACGCCTCCATGTCTACGCTTTTTTGGCATGTGCGGGCAGGAAGTAGGGGAAGACCGGCCCTCGGCGCGACGGGCCCGAAAAAAGGA
                                                  ‾‾‾‾‾‾‾‾‾‾‾‾‾‾‾‾‾‾‾‾‾‾‾‾‾‾‾‾‾‾‾‾‾‾‾‾‾‾‾‾

 50.        60        70        80        90       100       110       120       130       140
‾‾CCGGGGGCGATTCCCGCCTCGGTCCCCGGTCTCGGGAAGCTCCGCGGTGAGTCTCGCTCCCCGGCCCGATGATCTGCAACCCGCGCCCGGG→
                                                                                     ▼
```

Figure 2. Nucleotide sequence of Xenopus laevis ribosomal DNA (pX1rl4) surrounding the site of transcription initiation. The noncoding strand of the DNA is shown. Underlined nucleotides correspond to the 5' end of the 7.5 kb rRNA precursor. Triangles indicate cleavage sites for SmaI.

Transcription of Ribosomal DNA In Vitro

The molecular dissection of rDNA transcription controls requires an in vitro system where polymerase I will initiate accurately. Previous efforts to achieve this goal have only succeeded
in demonstrating various combinations of polymerase and template
that are not sufficient to yield accurate transcription (27,28).
It has been especially disappointing that nuclei, isolated from a
variety of cell types, are in general unable to reinitiate rRNA
synthesis (29-31). A hopeful exception to this general observation is the case of the large nuclei from Xenopus oocytes. These
nuclei are known to store large excesses of many nucleic acid-
related enzymes including RNA polymerases (21,32-34). As mentioned
above, they also contain a large amount of transcriptionally active
amplified rDNA. These nuclei can be manually isolated and nuclear
homogenates will support rRNA synthesis for up to 8 hr using the
endogenous amplified rDNA as template (35; Hipskind and Reeder,
unpublished data). Much of this synthesis is due to continued and
accurate reinitiation by polymerase I as shown by the fact that
the system will incorporate γ-(thio)-nucleoside triphosphates (36)
at the 5' end of growing precursor rRNA molecules (35; Hipskind
and Reeder, unpublished data). Electron microscopy of the in vitro
reaction mixtures after varying times of incubation also supports
the conclusion that RNA polymerase is reinitiating (McKnight,
Hipskind and Reeder, unpublished data).

Trendelenburg and Gurdon (37) have recently shown that cloned
rDNA can be accurately transcribed after injection into living
oocyte nuclei. Only a small fraction of the injected genes was
observed to be transcribed after spreading for electron microscopy.
But of that small number of active genes, about half were tightly
packed with RNA polymerase in a manner suggesting that accurate
initiation and termination were occurring. They conclude from these
data that it is not the polymerase itself that is limiting but some
other factor that opens the promoter and which is in short supply
(37). This agrees with the biochemical observation that, so far,
addition of cloned rDNA to oocyte nuclear homogenates has not
resulted in any detectable stimulation of polymerase I activity.
The fact that extensive initiation is occurring in vitro on the
endogenous rDNA suggests that the promoter-opening factor is
limiting and largely bound to endogenous template. Once a way is
found to make this factor move over and function on exogenously-
added cloned rDNA the way will be open to begin the molecular
dissection of the rDNA transcriptional machinery.

Function of the Nontranscribed Spacer

The possible function(s) of the nontranscribed spacer has been
a puzzle since its existence was first described. Obviously, some

of this region must contain regulatory sequences such as promoters, terminators, and possibly DNA replication origins. But the available evidence so far suggests that these will turn out to be relatively short sequences needing, at most, a few hundred nucleotides to contain them all. Of what use, then, is the bulk of the spacer which: 1) is a repetitive simple sequence, 2) is of variable length and sequence even within an interbreeding population, and 3) evolves rapidly between closely related species? The most economical explanation is that spacers arise as a by-product of unequal crossing-over that presumably goes on within the ribosomal gene locus. Wellauer et al. (13) and Fedoroff and Brown (38) have argued that unequal crossing-over coupled with strong selection on the transcribed gene sequences could well account for the observed arrangement. Smith (39) has shown from computer simulation studies that in the absence of selection, unequal crossing-over will cause any sequence to degenerate into a repetitive sequence. This repetitive sequence may in turn foster the crossing-over process itself. On a slightly larger scale, unequal crossing-over also seems to account for the remarkable homogeneity of the conserved gene regions in rDNA (13,38,39).

We should also remain alert to the fact that the rDNA locus is subject to a number of as yet poorly understood biological regulations. These include nucleolar dominance (26), chromosome pairing (40,41) and amplification (42,43) among others. It is possible that spacers will also be important for some of these processes.

5S DNA

In addition to one each of the large 18S and 28S rRNAs that have been derived from the 7.5 kb rRNA precursor, each eukaryotic ribosome contains one molecule of 5S rRNA. One might have imagined, a priori, that since equimolar amounts of rDNA and 5S DNA gene product are required, the biology of both types of genes would be very similar. In fact, frogs employ quite different strategies for 5S DNA than they use for rDNA. During oogenesis when rDNA undergoes massive amplification, an adequate supply of 5S RNA is insured by turning on an additional family of 5S genes termed oocyte 5S DNA.

The existence of two 5S gene classes was first detected because of sequence differences in their RNA products. Somatic 5S genes produce a homogeneous population of RNA molecules while the oocyte-type 5S genes code for several sequence variants (44,45).

Despite this sequence variation, both oocyte and somatic 5S genes are expressed in the oocyte and both are incorporated into ribosomes that are stored for eventual use during early embryogenesis. Whether or not this ribosome heterogeneity has any developmental significance is an intriguing but unanswered question.

After fertilization, the oocyte 5S genes appear to be completely
silent (45,46).

In laevis, there are about 24,000 copies of oocyte 5S genes
and they are located at the telomeres of most (if not all) of the
chromosomes (47,48). The exact number and location of the somatic
5S genes is not yet known.

Representative oocyte and somatic 5S DNA from both laevis and
borealis have been cloned (49,50; D. Brown, J. Doering and R.
Peterson, personal communication) and examples of oocyte 5S DNA
from both species have been completely sequenced (38,51,52; Korn
and Brown, unpublished data).

Xenopus laevis Oocyte 5S DNA

A diagram outlining the structure of X. laevis oocyte 5S DNA
is shown in Figure 3a. The organization of this DNA has several
homologies with rDNA organization described above. Each repeating
unit contains a transcribed gene and nontranscribed spacer. The
repeating units are heterogeneous in length and different length
repeats can reside next to each other on the same chromosome
(49,53). The AT-rich region of the repeat (Figure 3a) located
to the left of the gene region is largely composed of multiple
repeats of a 15 nucleotide sequence. All of the length heterogeneity resides in this AT-rich region and is primarily due to
variations in the number of 15 nucleotide subrepeats present.
The complete nucleotide sequence of the example diagrammed in
Figure 3 is shown in Figure 4. Near the left-hand end of the AT-
rich spacer the 15 nucleotide subrepeats are all precisely the
same. To either side of this region, the precision of the subrepeat degenerates. It has been proposed (51) that the precise
subrepeats have arisen relatively recently while the degenerate
ones are older and have had time to accumulate mutations. For
laevis oocyte 5S DNA, as well as for rDNA, it appears that duplication events can involve either the small subrepeats in the nontranscribed spacer or the entire repeating unit. In fact, it is
likely that both types of duplication are the result of the same
crossing-over mechanism (13,51).

To the right of the transcribed gene (Figures 3 and 4) is a
region that has been called the pseudogene (54). The sequence
of this region suggests that it arose by duplication of a portion
of the true gene and its promoter followed by mutational divergence.
Pseudogene sequences are not observed in the 5S RNA of oocytes; if
these sequences are transcribed they are not stable.

The pseudogene appears to be an archaeological remnant of an
earlier duplication event. Once the promoter of the pseudogene
became inactivated, or its RNA product became unstable, it was
no longer subject to selective pressure and began to diverge from
the true gene. All of the copies of the pseudogene have diverged

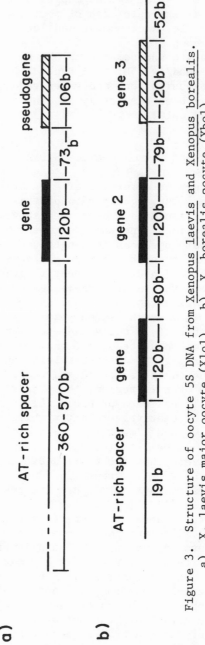

Figure 3.   Structure of oocyte 5S DNA from Xenopus laevis and Xenopus borealis.
        a) X. laevis major oocyte (X1o1). b) X. borealis oocyte (Xbo1).

Figure 4. Sequence of Xenopus laevis oocyte 5S DNA (Xlo1). The sequence shown is a continuous sequence read left to right and top to bottom. It has been arranged in this form to emphasize its repetitive nature. The noncoding strand of the DNA is shown.

in parallel. This can be explained if we assume that most recombi-
nation-duplication events in this gene cluster occur at the level
of the full repeating unit. Thus, the pseudogene could be passively
carried along and kept homogeneous by the same forces that maintain
the homogeneity of the true gene sequence (38).

Directly in front of the gene sequence is a region of about 50
nucleotides that almost certainly contains the promoter for the
gene. The sequence and function of this region will be considered
further on when we discuss the transcription of 5S genes.

A minor type of X. laevis oocyte 5S DNA which is present in
about 2000 copies per haploid genome has also been cloned (50).
In this DNA, the repeating unit is about 350 nucleotides long and
the transcribed gene region alternates with a GC-rich spacer whose
sequence is not related to that of the major oocyte 5S DNA. No
repetitive subrepeats have been detected in this spacer.

Xenopus borealis Oocyte 5S DNA

Representatives of the oocyte-type 5S genes of X. borealis
have been cloned (D. Brown, personal communication) and one fragment, designated Xbol, (Figure 3b) has been completely sequenced
(Figure 5) (Korn and Brown, unpublished data). Approximately
9000 of these genes are present per haploid genome (47). The
arrangement of gene and spacer in this DNA is considerably more
heterogeneous than is seen in laevis oocyte 5S DNA. A major
portion of the genes are arranged in small clusters varying
from 2 to 6 genes per cluster and separated from each other within
the cluster by a spacer of about 80 nucleotides. Separating the
gene clusters from each other are AT-rich spacers averaging several
thousand nucleotides in length (this region in Xbol is atypically
short). The AT-rich spacers differ in length. Part of this
variability is due to differing numbers of a 21 nucleotide subrepeat.

In borealis there are several different oocyte specific 5S
RNA sequences which differ both from laevis oocyte-type sequences
and from the sequence of borealis somatic 5S RNA (55). The cloned
fragment Xbol shown in Figure 5 contains three transcribed gene
regions. Gene 1 has the same sequence as does the major oocyte
type 5S RNA. The second gene differs from the first gene in two
positions, while the third gene differs from the first in fifteen
positions. RNAs transcribed from the second and third genes have
yet to be detected in borealis 5S RNA even though these two genes
appear to have active promoters (see below).

Somatic 5S Genes of laevis and borealis

The somatic 5S genes have been cloned from both frog species
and have been shown to have a tandemly linked gene-spacer arrangement.

```
                                                                 HhaIII               HhaI
  1                                  40                          |                    |         80
  AAGCTTCAAAAGTTATCTGTCTGAGCATCGCGTCGTTTTTGTCGCGCGTCGCGTCGTTTTTGTCGCGCGTCGCTTCGTTTTTTTTCGAGAATTTTCGCAAAGCGACAAATAATC

       120                                   160
  GTAGACTTTTGCAAGGTTAAAGTTTTGCACTTTTTTCGTCAAAGTCTTCATAGAAGCGTCAAAAGTCTTCACTCTGAT
Genes  HaeIII                                                     HpaII
  5'     200|                                240        HaeIII| |    |HaeIII               280                          3'
1 GCCTACGGCCACACCCCCCTGAAAGTGCCCGATCTCGTCTGATCTCGGAAGCGATGCAGGGCCGGGCCTGGTTAGTACCTGGATGGGAGACCGCCTGGGAATACCAGGTGTCGTAGGCTT

         320                                 360
  TTAGACTTTTGCCAGGTCAAAGTTTTGCAGGGTTTTTCGTCAAAGTCTTCATAGAAGCGTCAAAAGTCTTCACTCCGAT
       HaeIII                                                  HaeIII
  5'     400|                                440               |                        480                           3'
2 GCCTACGGCCACACCCCCCTGAAAGTGCCCGATCTCGTCTGATCTCGGAAGCGATGCAAGGTCGGGCCTGGTTAGTACCTGGATGGGAGACCGCCTGGGAATACCAGGTGTCGTAGGCTT

         520                                 560
  TTAGACTTTTGCCAGGTCAAAGTTTTGCAGGGTTTTTCTTCAAAGTCTTCATAGAAGCGTCAAAAGTCAGCAAACCTA
                                                            HaeIII
  5'      600                                640            |                            680                          3'
3 CCCTGCGGCTACACCCCCCTGAAAGAGCCTGATCTTGTCTGATCTCAGAAGCCATGCGGGGCCGACCCTGGTTAGTACCTGGATGGGAGACCACCTGGGAATACAAGGTGTTGTAGTCTT

          720                                760
  TTTCCAGGTCAAAATTTTGCAGGTTTTTTCTTCAAAGTCTTCATAGAAGCTT
```

Figure 5. Sequence of Xenopus borealis oocyte 5S DNA (Xbo1). This is a continuous sequence read left to right and top to bottom. The noncoding strand of the DNA is shown.

The somatic 5S DNA of _borealis_ has a homogeneous repeat length of
about 850 nucleotides (J. Doering, personal communication). The
somatic 5S DNA of _laevis_ also appears homogeneous with a repeat
length of about 850 nucleotides (R. Peterson, personal communication).

Transcription of 5S DNA

In living cells, 5S DNA is transcribed by RNA polymerase III,
a polymerase that is sensitive to high levels of α-amanitin (21).
Mature 5S RNA can be isolated with polyphosphate termini at their
5' end (56). This suggests that the primary transcript is not
processed at this end of the molecule. At the 3' end, molecules
have been detected that are longer than normal 5S RNA (57). However, no precursor product relationship has been demonstrated between these longer molecules and normal length 5S RNA. It seems
equally possible that they represent an occasional failure to
recognize the termination signal.

Study of 5S gene transcription has received a major impetus by
the recent development of _in vitro_ systems that are capable of
accurately synthesizing 5S RNA. In contrast to polymerase I and
II, it has been known for some time that polymerase III is capable
of reinitiating RNA chains in some types of isolated nuclei.
Parker and Roeder (58) extended this by isolating a crude chromatin
from immature _Xenopus_ oocytes (active in 5S synthesis) and showing
that addition of purified polymerase III caused a several-fold
stimulation in 5S RNA production. Similar stimulations of polymerase III gene products have since been obtained by addition of
purified polymerase III to several different types of isolated
nuclei (59).

More recently Wu (60) has developed a fully soluble system in
which a cytoplasmic extract from KB cells directs accurate transcription of genes recognized by polymerase III when added to
purified DNA (60; R. Roeder, personal communication). Polymerase
III and its cofactor(s) are rapidly lost from the nucleus during
cell lysis thus accounting for their anomalous presence in the
cytoplasm. It appears likely that this system will soon be fractionated and the molecular components required for 5S gene
transcription will be isolated and characterized.

In parallel with the above work, the _Xenopus_ oocyte nucleus
has been developed as a highly useful tool for studying 5S gene
transcription. It was first shown that injection of uncloned
genomic 5S DNA into the oocyte nucleus resulted in a massive
stimulation of accurate 5S RNA synthesis (61). This was later
extended by showing that cloned single repeating units of 5S DNA
could yield the same result (62). The transcription was α-amanitin
sensitive indicative of polymerase III activity and the type of
RNA made was completely dependent upon the coding capacity of the
DNA injected. This was the first demonstration that genes amplified

via the molecular cloning process still retained all the signals
necessary for accurate transcription. It also opened the door for
experiments in which in vitro mutagenesis of cloned 5S genes could
be used to define precisely which nucleotides are involved in promoter function.

Extending this approach the next logical step, Birkenmeier,
Brown and Jordan (63) have shown that homogenates of manually
isolated oocyte nuclei can also direct accurate 5S RNA synthesis.
Addition of cloned 5S DNA can stimulate 5S RNA synthesis in
homogenates by at least 10-fold. This suggests that in oocytes,
the promoter-opening factor(s) for 5S DNA are not as limiting as
they appear to be for rDNA (see above) and that they are more
freely accessible to exogenously added genes.

The availability of transcription systems that will accurately
transcribe cloned 5S DNA makes it possible to begin defining
which nucleotides are involved in RNA polymerase initiation and
termination. One approach to this question is to compare the
flanking sequences of all genes that have been shown to be transcribed by polymerase III and look for regions of homology. Korn
and Brown (unpublished data) have done this comparison for the 5'
flanking sequences of six eukaryotic genes (Figure 6) and also for
the 3' flanking sequences of the same six genes (Figure 7). The sequences of Figures 6 and 7 have been shown to be transcribed by
oocyte nuclear homogenates.

A computer search (64) of these sequences reveals the following homologies. At the 5' end is found: GAC, then 5 or 6 residues,
AGAAG, then 3 or 6 residues, AAAAG, then 13 or 14 residues, transcription initiation. The significance of these homologies is
enhanced by the relative invariance of their position with respect
to transcription initiation. Yeast 5S DNA showed poor homology to
the above sequence, in agreement with the observation that yeast
polymerase III does not appear to recognize Xenopus 5S DNA promoters (R. Roeder, personal communication). The spacing between
the conserved sequences GAC, AGAAG and AAAAG is about 10 base pairs.
This is the spacing required to place all of these sequences on one
face of the DNA double helix. No significant dyad symmetries were
found in this region.

There is an abrupt change in nucleotide sequence in the oocyte
5S DNAs of both laevis (Figure 4) and borealis (Figure 5) that
occurs at about 50 nucleotides before the 5' end of the gene.
This suggests that this 50 nucleotides contains all the information
needed for accurate initiation of transcription. This suggestion
is strengthened by the demonstration that the entire AT-rich spacer
can be deleted and accurate transcription of the gene will still
occur (N. Fedoroff, personal communication). Furthermore, the
distance between genes 1 and 2 (79 nucleotides) and genes 2 and 3
(78 nucleotides) in Xbol (Figure 5) demonstrate that this amount
of DNA is sufficient to code for both termination of one gene and
initiation of the next.

```
                              -70        -60        -50        -40        -30        -20        -10        -1

X. borealis oocyte 5S    TTGCAAGGTTAAAGTTTTGCACTTTTTTCGTCAAAGTCTTCATAGAAGCGTCAAAAGTCTTCACTCTGAT

X. borealis somatic 5S   CCTGGCATGGGGAGGAGCTGGCCGCCCCCAGAAGGGGAGGAAAAGTCAGCCTTGTGCC

X. laevis oocyte 5S      AAAGTTTTCATTTTCATTTTCCACAGTGCCGCTGACAAGTCAAGAAGCCGAAAAGTGCCGCTGTTCATC

Adenovirus VA RNA_I      GGACGCCTCTGGCCGGTGAGGCGTGCGCAGTCGTTGACGCTCTAGACCGTGCAAAAGAGAGCCTGATAAGC

Drosophila 5S            CAGTCTATTTCAGTCTATGGGCATAACTGAATATCAGAGTATAAGGACACTGTTTAGCCCCTCGACTTTC

Yeast 5S                 CCTCTCACTCCCACCTACTGAACATGTCTGGAGCCTGCCCTCATATCACCTGCGTTTCCGTTAAACTATC
```

Figure 6. Comparison of nucleotide sequences preceding six eukaryotic genes transcribed by polymerase III. Homologies with GAC, AGAAG and AAAAG (see text) are in boxes. The homology found in yeast 5S DNA was not statistically significant.

GENE

X. borealis oocyte 5S

GGTTAGTACCTGGATGGGAGAGACCGCCTGGGAATACCAGGTGTCTCGTAGGCTTTAGACTTTTGCCAGGTCAAAGTTTGCAG

X. borealis somatic 5S

GGTTAGTACTTGGATGGGAGAGACCGCCTGGGAATACCAGGTGTCTCGTAGGCTTTTGCACTTTTGCCATTCTGAGTAACAGCAG

X. laevis oocyte 5S

GGTTAGTACCTGGATGGGAGAGACCGCCTGGGAATACCAGGTGTCTCGTAGGCTTTCAAAGTTTCAACTTTATTTTGCCACAG

Adenovirus VA RNA_I

CCGCGTGTCGAACCCAGGTGTGCCGACGTCAGACAACGGGGAGGCGCTCCTTTGGCTTCCTTCCAGGCGCGGGGCCGCTGCTG

Drosophila 5S

GATGGGGGACCGCTTGGGAACACCGCGTGTTGTGGCCTCGTCCACAACTTTTTGCTGCCTGCCTCGCCTGCCTGCTGCTGCC

Yeast 5S

ACCGAGTAGTGTAGTGGGTGACCATACCGCAAACTCAGGTGCTGCAATCTTTATTTCTTTTTTTTTTTTTTTTTTTTTTT

Figure 7. Comparison of the terminal portion and 3' flanking sequences for six eukaryotic genes transcribed by polymerase III. The sequences are aligned with respect to the first T residue of the T cluster at which termination occurs. The two arms of dyad symmetry regions are underlined by arrows.

Although homologies have been found in the putative promoters of all 5S genes so far examined, it is clear that some variation in promoter sequence is still compatible with correct initiation. Studies with the oocyte nuclear homogenate system have shown that there is a wide range of promoter strength among the various genes tested. For example, at similar DNA input laevis somatic 5S DNA (whose sequence is not yet determined) is about 50 times more efficient than laevis oocyte 5S DNA in promoting RNA synthesis (R. Peterson and E. Birkenmeier, personal communication).

Figure 7 compares the 3' flanking sequences of six eukaryotic genes transcribed by polymerase III. Korn and Brown (26) conclude that for all six genes the sequences preceding termination are GC-rich (possibly to slow down transcription) while termination occurs in an AT-rich region containing a string of T's. In laevis oocyte 5S DNA, a second T cluster, 10 nucleotides removed from the normal termination site, can also function if the first stop signal is missed (57).

$$tDNA_1^{met}$$

A third class of multiple copy genes that have been studied in X. laevis are the genes coding for amino acid acceptor tRNA. In particular a fraction of DNA coding for the initiation methionyl tRNA ($tRNA_1^{met}$) has been partially purified by density gradient centrifugation (65). The $tRNA_1^{met}$ coding regions are clustered and are reiterated about 300-fold (66,67). Several restriction enzymes (EcoRI, HapI, HindIII) cut $tDNA_1^{met}$ at regular intervals of 3.18 kb. No evidence has been found so far to indicate any heterogeneity in this repeat length or in any simple sequence repetitive DNA.

Several repeat length HindIII fragments of $tDNA_1^{met}$ have been cloned (68) utilizing the single HindIII site present in the phage vector λ 598 (69). A battery of restriction enzyme sites have been mapped on this 3.18 kb fragment. The data suggest the structure shown diagrammatically in Figure 8 (68). The fragment contains 2 $tRNA_1^{met}$ genes, both in the same orientation, and separated from each other by a distance of 0.35 kb. The genes were located by hybridizing various restriction fragments with labeled $tRNA_1^{met}$ and by the fact that only two regions of the 3.8 kb repeat contain a cluster of 5 restriction sites in the proper order and spacing as predicted from the known sequence of $tRNA_1^{met}$ (70). No evidence has yet been found for interrupting sequences such as are present in some yeast tRNA genes (71,72).

In addition to the two genes for $tRNA_1^{met}$, the 3.18 kb repeat contains at least one other tRNA gene of unknown type located within the right half of the repeat (see Figure 8). No other coding sequences have been detected, even after saturating the $tDNA_1^{met}$ with long term labeled total cellular RNA from X. laevis cultured cells. Thus, it seems likely that most of this repeating unit is nontranscribed spacer.

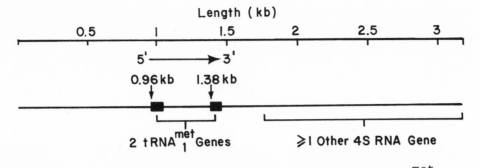

Figure 8. Diagram of the structure of <u>Xenopus</u> <u>laevis</u> tDNA$_1^{met}$.

In the living cell, tDNA$_1^{met}$ is also transcribed by RNA polymerase III. Injection of cloned tDNA$_1^{met}$ into <u>Xenopus</u> oocyte
nuclei indicates that at least two of the tRNA genes on the 3.18
kb fragment are transcriptionally active (73). Present evidence
suggests that this cloned repeat does not harbor a nonfunctional
pseudogene, as is present in <u>X</u>. <u>laevis</u> oocyte 5S DNA, but defini-
tive proof of this must await further work.

VITELLOGENIN GENES

A fourth class of genes that are just beginning to be studied
in <u>Xenopus</u> are those coding for vitellogenin, the precursor to
yolk protein. In adult frogs, the livers of females normally
secrete vitellogenin into the blood where it is taken up by developing oocytes in the ovary. Ordinarily, male livers do not
synthesize this protein. Upon injection with estrogen, however,
they can be induced to synthesize large amounts of the protein.
Tadpole livers, in contrast, do not respond to the hormone.
The vitellogenin mRNA from <u>X</u>. <u>laevis</u> has been characterized
(74,75) and its hormone-dependent accumulation has been measured
(75,76). Purified vitellogenin mRNA behaves as a single component
with a homogeneous length of 6.3 kb and Wahli et al. (77) have
prepared a number of cDNAs cloned from the mRNA. Despite the
size homogeneity of the mRNA, comparison of these cDNA clones
with each other and against purified mRNA has revealed an interesting heterogeneity. Contrary to conclusions made from earlier
kinetic complexity analyses (74), vitellogenin is coded not by a
single gene, but by a small family of genes. The cDNAs fall into
two major groups that differ from each other by about 20% of their
nucleotides. Each of these major groups can be divided into at
least two subgroups that differ from each other by about 5%.

Hybridization of cloned cDNA to restriction enzyme fragments of
total genomic DNA shows that both major groups are present in the
DNA of a single animal. It is likely that both major groups are
expressed as RNA in each animal.

Hybridization to restricted genomic DNA also suggests that the
genes contain intervening sequences not present in the mature mRNAs.

REFERENCES

1   Reeder, R.H. (1974) in The Ribosome (Nomura, M., Tissières, A.
    and Lengyel, P., eds.), pp. 489–518, Cold Spring Harbor Labora-
    tory, Cold Spring Harbor, NY.
2   Birnstiel, M., Speirs, J., Purdom, I., Jones, K. and Loening,
    U.E. (1968) Nature 219, 454.
3   Dawid, I.B., Brown, D.D. and Reeder, R.H. (1970) J. Mol. Biol.
    51, 341–360.
4   Morrow, J.F., Cohen, S.N., Chang, A.C.Y., Boyer, H.W., Good-
    man, H.M. and Helling, R.B. (1974) Proc. Nat. Acad. Sci.
    U.S.A. 71, 1743–1747.
5   Wellauer, P.K. and Dawid, I.B. (1974) J. Mol. Biol. 97, 379–395.
6   Dawid, I.B. and Wellauer, P.K. (1976) Cell 8, 443–448.
7   Reeder, R.H., Higashinakagawa, T. and Miller, O. Jr. (1976)
    Cell 8, 449–454.
8   Boseley, P.G., Tuyns, A. and Birnstiel, M.L. (1978) Nucl. Acids
    Res. 5, 1121–1137.
9   Wellauer, P.K., Reeder, R.H., Carroll, D., Brown, D.D.,
    Deutch, A., Higashinakagawa, T. and Dawid, I.B. (1974) Proc.
    Nat. Acad. Sci. U.S.A. 71, 2823–2827.
10 Wellauer, P.K., Dawid, I.B., Brown, D.D. and Reeder, R.H. (1976)
    J. Mol. Biol. 105, 461–486.
11 Botchan, P., Reeder, R.H. and Dawid, I.B. (1977) Cell 11, 599–
    607.
12 Brown, D.D., Dawid, I.B. and Reeder, R.H. (1977) Develop. Biol.
    59, 266–267.
13 Wellauer, P.K., Reeder, R.H., Dawid, I.B. and Brown, D.D. (1976)
    J. Mol. Biol. 105, 487–505.
14 Buongiorno-Nardelli, M., Amaldi, F., Beccari, E. and Junakovic,
    N. (1977) J. Mol. Biol. 110, 105–117.
15 Kirkman, H.N. and Southern, E.M. (1978) Fed. Proc. 37, 1500A.
16 Bird, A.P. (1977) Cold Spring Harbor Symp. Quant. Biol. 42,
    1179–1183.
17 Reeder, R.H., Brown, D.D., Wellauer, P.K. and Dawid, I.B. (1976)
    J. Mol. Biol. 105, 507–516.
18 Scheer, U., Trendelenburg, M.F., Krohne, G. and Franke, W.W.
    (1977) Chromosoma 60, 147–167.
19 Bird, A.P. and Southern, E.M. (1978) J. Mol. Biol. 118, 27–47.
20 Bird, A.P. (1978) J. Mol. Biol. 118, 49–60.

21 Roeder, R.G. (1976) RNA Polymerase (Losick R. and Chamberlin
    M., eds.), p. 285, Cold Spring Harbor Laboratory, Cold Spring
    Harbor, NY.
22 Landesman, R. and Gross, P.R. (1968) Develop. Biol. 18, 571–589.
23 Miller, O.L. and Beatty, B.R. (1969) Science 164, 955–957.
24 Reeder, R.H., Sollner-Webb, B. and Wahn, H.L. (1977) Proc.
    Nat. Acad. Sci. U.S.A. 74, 5402–5406.
25 Martin, S.A. and Moss, B. (1975) J. Biol. Chem. 250, 9330–9335.
26 Korn, L. and Brown, D.D., Cell (in press).
27 Roeder, R.G., Reeder, R.H. and Brown, D.D. (1970) Cold Spring
    Harbor Symp. Quant. Biol. 35, 727–735.
28 Honjo, T. and Reeder, R.H. (1974) Biochemistry 13, 1896–1899.
29 Zylber, E.A. and Penman, S. (1971) Proc. Nat. Acad. Sci.
    U.S.A. 68, 2861–2865.
30 Grummt, I. and Lindigkeit, R. (1973) Eur. J. Biochem. 36,
    244–249.
31 Ferencz, A. and Seifart, K.H. (1975) Eur. J. Biochem. 53,
    605–613.
32 Benbow, R.M., Pestell, R.Q.W. and Ford, C.C. (1975) Develop.
    Biol. 43, 159–174.
33 Woodland, H.R. and Adamson, E.D. (1977) Develop. Biol. 57,
    118–135.
34 Laskey, R.A., Mills, A.D. and Morris, N.R. (1977) Cell 10,
    237–243.
35 Reeder, R.H., Sollner-Webb, B., Hipskind, R., Wahn, H.L. and
    Botchan, P. (1978) 10th Annual Miami Winter Symp. (in press).
36 Reeve, A.E., Smith, M.M., Pigiet, V. and Huang, R.C.C. (1977)
    Biochemistry 16, 4464–4469.
37 Trendelenburg, M.F. and Gurdon, J.B. (1978) Nature (in press).
38 Fedoroff, N.V. and Brown, D.D. (1977) Cold Spring Harbor Symp.
    Quant. Biol. 42, 1195–1200.
39 Smith, G.P. (1976) Science 191, 528–535.
40 McClintock, B. (1934) Z. Zellforsch. Mikroskop. Anat. Abt.
    Histochem. 21, 294–328.
41 Evans, M.J., Buckland, R.A. and Pardue, M.L. (1974) Chromosoma
    48, 405–426.
42 Brown, D.D. and Dawid, I.B. (1968) Science 160, 272–280.
43 Tobler, H. (1975) Biochemistry of Animal Development, Vol. III,
    pp. 91–143, Academic Press, New York, NY.
44 Wegnez, M., Monier, R. and Denis, H. (1972) FEBS Letters 25, 13.
45 Ford, P.J. and Southern, E.M. (1973) Nature New Biol. 241, 7–12.
46 Miller, L. (1974) Cell 3, 275–281.
47 Brown, D.D. and Sugimoto, K. (1973) J. Mol. Biol. 78, 397–415.
48 Pardue, M.L., Brown, D.D. and Birnstiel, M.L. (1973) Chromosoma
    42, 191–203.
49 Carroll, D. and Brown, D.D. (1976) Cell 7, 477–486.
50 Brown, D.D., Carroll, D. and Brown, R.D. (1977) Cell 12, 1045–
    1056.
51 Fedoroff, N.V. and Brown, D.D. (1978) Cell 13, 701–716.

52   Miller, J.R., Cartwright, E.M., Brownlee, G.G., Fedoroff, N.V.
     and Brown, D.D. (1978) Cell 13, 717–725.
53   Carroll, D. and Brown, D.D. (1976) Cell 7, 467–475.
54   Jacq, C., Miller, R.J. and Brownlee, G.G. (1977) Cell 12, 109–
     120.
55   Ford, P.J. and Brown, R.D. (1976) Cell 8, 485–493.
56   Monier, R. (1974) in The Ribosome (Nomura, M., Tissières, A.
     and Lengyel, P., eds.) p. 141, Cold Spring Harbor Laboratory,
     Cold Spring Harbor, NY.
57   Denis, H. and Wegnez, M. (1973) Biochemie 55, 1137–1151.
58   Parker, C.S. and Roeder, R.G. (1977) Proc. Nat. Acad. Sci. U.S.A.
     74, 44–48.
59   Parker, C.S., Jaehning, J.A. and Roeder, R.G. (1977) Cold
     Spring Harbor Symp. Quant. Biol. 42, 577–587.
60   Wu, G. (1978) Proc. Nat. Acad. Sci. U.S.A. 75, 2175–2179.
61   Brown, D.D. and Gurdon, J.B. (1977) Proc. Nat. Acad. Sci. U.S.A.
     74, 2064–2068.
62   Brown, D.D. and Gurdon, J.B. (1978) Proc. Nat. Acad. Sci. U.S.A.
     75, 2849–4853.
63   Birkenmeier, E., Brown, D.D. and Jordan, E. Cell (in press).
64   Korn, L.J., Queen, C.L. and Wegman, M.N. (1977) Proc. Nat.
     Acad. Sci. U.S.A. 74, 4401–4405.
65   Clarkson, S.G. and Kurer, V. (1976) Cell 8, 183–195.
66   Clarkson, S.G., Birnstiel, M.L. and Purdom, I.F. (1973) J. Mol.
     Biol. 79, 411–429.
67   Clarkson, S.G., Birnstiel, M.L. and Serra, U. (1973) J. Mol.
     Biol. 79, 391–410.
68   Clarkson, S.G., Kurer, V. and Smith, H.O. (1978) Cell 14
     713–724.
69   Murray, N.E., Brammar, W.J. and Murray, K. (1977) Mol. Gen.
     Genet. 150, 53–61.
70   Wegnez, M., Mazabrand, A., Denis, H., Petrissant, G. and
     Boisnard, M. (1975) Eur. J. Biochem. 60, 295–302.
71   Goodman, H.M., Olson, M.V. and Hall, B.D. (1977) Proc. Nat.
     Acad. Sci. U.S.A. 74, 5453–5457.
72   Valenzuela, P., Venegas, A., Weinberg, F., Bishop, R. and Rutter,
     W.J. (1978) Proc. Nat. Acad. Sci. U.S.A. 75, 190–194.
73   Kressman, A., Clarkson, S.G., Pirotta, V. and Birnstiel, M.L.
     (1978) Proc. Nat. Acad. Sci. U.S.A. 75, 1176–1180.
74   Wahli, W., Wyler, T., Weber, R. and Ryffel, G.U. (1976) Eur.
     J. Biochem. 66, 457–465.
75   Shapiro, D.J. and Baker, H.J. (1977) J. Biol. Chem. 252,
     5244–5250.
76   Ryffel, G.U., Wahli, W. and Weber, T. (1977) Cell 11, 213–221.
77   Wahli, W., Ryffel, G.U., Wyler, T., Jaggi, R.B., Weber, R.
     and Dawid, I.B. (1978) Develop. Biol. (in press).

TRANSFORMATION OF YEAST

Christine Ilgen, P.J. Farabaugh, A. Hinnen,
Jean M. Walsh and G.R. Fink

Section of Botany, Genetics and Development
Cornell University
Ithaca, New York 14853

INTRODUCTION

Yeast transformation, recently described by Hinnen et al. (1), permits the introduction of cloned DNA segments into the genome of the yeast Saccharomyces cerevisiae. The system requires a highly enriched source of yeast DNA and a reasonably stable recipient yeast strain.

There are a number of ways to enrich for a desired yeast DNA sequence. The total yeast genome can be represented by a single collection of a few thousand E. coli strains, each of which contains a plasmid with a yeast DNA insert. Specific yeast genes in these E. coli collections have been identified biochemically, by nucleic acid hybridization, or functionally by genetic complementation of equivalent E. coli mutations. Radioactive RNA probes, isolated and purified from yeast, have been utilized for screening such E. coli clone banks by the Grunstein-Hogness colony hybridization procedure (4). The hybridization approach has been used to identify within E. coli banks yeast sequences for which there is an abundant RNA transcript: tRNA genes (5), ribosomal RNA genes (6), glycolytic genes (7), and highly inducible genes (galactose) (M. Schell and D. Wilson, personal communication). Similarly, a clone containing the yeast iso-1-cytochrome c gene has been identified with a DNA hybridization probe synthesized chemically to correspond to the DNA sequence inferred from mutationally altered protein sequences (8). Some yeast genes, when present on hybrid plasmids, will complement mutations in the homologous genes of E. coli. The complementation approach has worked for the leu2, his3, trp1, arg4 and trp5 genes of yeast which complement the leuB, hisB, trpC, argH and trpA genes of E. coli, respectively (9-11). However, in one

117

study, only about 20% of the yeast genes tested actually complemented the homologous E. coli mutations (11). Those genes that are
not expressed in E. coli or fail to produce abundant transcripts
can be identified by transformation from the E. coli bank into
yeast. Yeast transformation has the advantage that the function
of the cloned yeast gene is assayed in vivo in its natural host.
As in any transformation system, the use of a stable recipient
permits the detection of rare transformation events.

Two basic types of transformation events have been uncovered
in yeast: an integrative type, in which the transforming DNA becomes associated with a yeast chromosome, and a nonintegrative
type, in which at least a portion of the transforming DNA is maintained as a plasmid. The type of transformation event is determined
by the specific yeast sequences used as the transforming DNA. Many
yeast genes, such as leu2, his3, his4 or ura3, transform at a low
frequency in integrative-type transformation reactions (1,3; D.
Botstein, personal communication; P. Farabaugh, unpublished data).
Certain yeast genes, trp1, arg4 and ribosomal DNA, for example,
transform yeast at frequencies 100- to 1000-fold greater than other
yeast genes and remain in the yeast cell as autonomously replicating,
extrachromosomal elements (2; J. Carbon, personal communication;
J. Szostak, personal communication). Genes associated with the
endogenous yeast plasmid, 2-micron-circular (2μ) DNA, also transform yeast cells at high frequency (2,3). An unusual feature of
cells transformed with 2-micron DNA is that they contain an integrated copy of the plasmid in addition to several unintegrated
copies, which are retained as plasmids (2). Integrative-type transformants are generally stable for the transformed phenotype, while
transformants of the nonintegrative type are highly unstable. When
genes that normally integrate during transformation are associated
with trp1, rDNA or 2-micron-circular DNA, they transform at high
frequency and fail to integrate. Presumably, the trp1, rDNA and
2-micron segments contain sequences that allow autonomous replication.

YEAST TRANSFORMATION PROTOCOL

Yeast transformation is carried out according to the published
method of Hinnen et al. (1). This procedure is based on the co-
precipitation of calcium-treated yeast spheroplasts and transforming
DNA by polyethylene glycol 4000 (PEG 4000). The spheroplasts are
then embedded into regeneration agar and plated on selective medium.

Source of DNA

Yeast DNA, inserted into a standard cloning vector such as the
E. coli plasmids ColE1 or pBR322, bacteriophage lambda or the yeast
2μ DNA, is used to transform yeast cells. The existence of foreign
DNA on the vector enables a yeast transformation event, first

identified by growth of the recipient on selective medium, then
verified by the biochemical detection of vector DNA sequences
within transformed cells. Through the use of the Southern hybridization procedure (12) as well as a colony hybridization technique
developed for yeast (see below), vector sequences can be identified
within the yeast genome and their segregation monitored in genetic
crosses.

Preparation of Spheroplasts

An overnight culture of yeast is inoculated into rich medium,
such as YEPD (1% yeast extract, 2% bacto-peptone and 2% dextrose),
and the cells are grown to a density of 2 to 3 x 10^7 cells/ml.
The cells are washed once with 1 M sorbitol and resuspended in 1 M
sorbitol at a concentration of 2 to 3 x 10^8 cells/ml. Glusulase
(Endo Labs, Garden City, NY) is added to a final concentration of
1% and the resulting suspension is then incubated with gentle
shaking at 30° for 1 hr.

Transformation

Following glusulase treatment, the spheroplasts are washed
twice with 1 M sorbitol by low speed centrifugation at room temperature. After the second wash, the spheroplasts are taken up
in a solution containing 1 M sorbitol, 10 mM $CaCl_2$ and 10 mM Tris-
HCl, pH 7.5, spun again and resuspended in the same solution at a
concentration of 2 to 3 x 10^9 cells/ml. The DNA is added at this
point and the suspension incubated for 20 to 30 min at room temperature. Then a 10-fold volume of 44% PEG is added to give a
final PEG concentration of 40% and the spheroplast/DNA aggregate
is incubated for 10 min at room temperature. Finally, the spheroplasts are centrifuged again, resuspended in 1 M sorbitol and added
to warmed (46 to 48°C) regeneration agar (3% agar in 1 M sorbitol
and an enriched, selective medium). Samples of 10 ml each are
plated immediately onto selective agar plates (approximately 2 to
3 x 10^8 cells/plate).

Transformed spheroplasts generally give rise to visible
colonies within three days. With some DNAs, such as those carrying
leu2 and his4 genes, both large and small colonies appear on the
plates (Figure 1). Large colonies are composed of stably transformed cells. Most microcolonies do not grow beyond 0.5 mm in
diameter and rapidly lose the transformed phenotype upon restreaking. Occasional microcolonies eventually become large;
these contain stably transformed cells indistinguishable from
normal transformants. Microcolonies may represent a transformation event in which transforming DNA exists in the cell as a
transient plasmid. Stable transformation is subsequently achieved

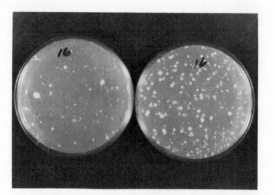

Figure 1. Transformation of his4 yeast strain with a hybrid plasmid (pYehis4) carrying the his4 genes of yeast. The plasmid consists of a BamHI fragment of yeast inserted into the BamHI site of
pBR313. Transformed yeast cells are apparent as colonies embedded
in the agar. The plate on the left illustrates the transformation
frequency obtained with 0.2 μg of intact plasmid DNA. The plate on
the right demonstrates the increase in transformation efficiency
obtained with an equivalent amount of pYehis4 that has been digested
with BamHI.

in a small percentage of these transformants by integration of plasmid sequences into the chromosomes.
     Usually, the transformants of the nonintegrative type give
rise to small colonies. It is not clear whether the small size is
related to instability of the unintegrated plasmid or to inefficient expression of the gene on the plasmid.

                          General Comments

     A good spheroplast preparation is crucial to the success of
transformation. The time needed to digest cell walls with glusulase is dependent on the age of the cells, with longer incubation
periods required for older cells. The effectiveness of the glusulase treatment in spheroplast formation can be monitored by
determining the viable cell count on standard YEPD agar plates
(without sorbitol) before and after treatment with glusulase.
Spheroplasts are osmotically sensitive and will not regenerate on
these YEPD plates. Hence, the viable cell count after glusulase
treatment should drop to 0.1 to 1% of the pretreatment titer on
these plates.
     The DNA can be added to the spheroplasts in a low salt buffer
if the molarity of the sorbitol is not reduced to less than about
0.8 M. Circular as well as linear DNA can be used, and the transformation frequency of circular DNA is directly proportional to

DNA concentration, within the range from 1 to 100 µg DNA/ml. It is
advisable to have the selective medium as rich in nutrients as
possible since the regeneration of spheroplasts works best in a
complex medium. In transformation experiments for amino acid
markers, the yeast nitrogen base minimal medium is routinely
enriched with 2% YEPD and a number of amino acids.

ANALYSIS OF YEAST TRANSFORMANTS

Transformed yeast strains are purified by restreaking on
selective media and are then analyzed further.
Genes introduced by transformation of haploid strains can be
followed through appropriate crosses using the standard yeast
genetic procedures described by Sherman et al. (13). A particularly useful feature of the yeast life cycle is the retention of
the four haploid products of meiosis within an ascus wall as a
tetrad. By tetrad analysis (separation of the spores by micro-
manipulation and subsequent phenotypic analysis of the isolated
spores (14)), the segregation of transforming DNA sequences can
be followed. Tetrad analysis has been fully described by Mortimer
and Hawthorne (15).

Southern Analysis

If transforming DNA carries foreign sequences such as E. coli
plasmid DNA, the foreign DNA can be utilized in nucleic acid hybridization as a probe to detect transforming DNA within transformed
cells. In the Southern blotting technique (12) DNA from an agarose
gel is transferred to a nitrocellulose filter and hybridized to a
radioactively labeled probe. Nonintegrated transforming DNA exists
as a small supercoiled plasmid and thus migrates differently than
chromosomal DNA during agarose gel electrophoresis; its presence
can be shown easily using the Southern technique. In integrative
type transformation events, the local structure of the genome may
be altered by the introduction of foreign DNA sequences (e.g., plasmid DNA) along with yeast genes. Transposition of yeast genes also
may take place if the transforming yeast sequences integrate at
sites other than the normal chromosomal location. In either case,
Southern analysis reveals the alteration in the restriction endonu-
cleolytic pattern generated by the insertion of the transforming DNA.

Colony Hybridization

Yeast colony hybridization, based on the Grunstein-Hogness
filter hybridization technique for bacterial colonies (4), allows
easy verification of the transformation event and rapid characterization of large numbers of yeast transformants. Yeast colonies are

grown, spheroplasts generated and lysed, and DNA denatured and
fixed in situ on a nitrocellulose filter. Radioactivity labeled
DNA or RNA probes are then employed to screen the colonies for
specific sequences present in DNA fixed to the filter. Up to 50
colonies can be screened on a filter measuring 60 x 75 mm.

Yeast cells are spotted with a toothpick on a Millipore HA
nitrocellulose filter placed on either an enriched or selective
medium. Following overnight incubation at 30°, the filter is removed to a piece of blotting paper (Whatman 470) saturated with
.35 M β-mercaptoethanol in 50 mM EDTA, pH 9, and incubated for
15 min. After the filter dries, it is wet again with a solution
containing 1 M sorbitol in 50 mM EDTA, pH 7.5, dried and then
transferred to a petri dish containing a blotting paper saturated
with zymolyase (60,000 units per g, Kirin Brewery) at 0.1 mg/ml
in the sorbitol-EDTA solution. The zymolyase reaction is allowed
to proceed at 37° until at least 20% of the cells are present as
spheroplasts; this reaction usually takes 2 to 3 hr. Spheroplast
generation can be monitored by transferring a small sample of yeast
cells from the filter into denaturation buffer (0.2 M NaOH, 0.6 M
NaCl) (12) and following lysis of spheroplasts by microscopic
examination. It is essential that the zymolyase step be carried
out in a moist chamber.

Following the zymolyase reaction, the filter is placed
successively for treatments of 5 min each on blotting paper saturated with denaturation buffer, neutralization buffer (1.5 M NaCl,
1 M Tris-HCl, pH 7.4) (12) and 2X SSC twice (0.3 M NaCl, .03 M
sodium citrate, pH 7.0). All of these manipulations are carried
out at room temperature and the filter is dried after each treatment. The filter is then baked for 1 to 2 hr at 80° in a vacuum
oven. At this point, the filter is ready for hybridization.
Alternatively, it can be wrapped in plastic film and stored indefinitely. Hybridization is carried out according to published
methods (16).

Hybridization of ^{32}P-labeled ColE1 to cells transformed with
the plasmid pYeleu10 (a hybrid plasmid containing the yeast leu2
gene inserted into the E. coli plasmid ColE1) has revealed that
80 to 90% of all yeast cells transformed with this plasmid have
ColE1 sequences stably integrated into their genome. When the
transformants are analyzed genetically, colony hybridization can
be used in conjunction with tetrad analysis to determine the linkage
relationship of ColE1 and LEU2^{+} sequences. Figure 2 demonstrates
a cross between a pYeleu10 transformant and an untransformed leu2^{-}
strain; in each of the 10 tetrads shown, segregation of the Leu^{+}
phenotype is in the parental arrangement with ColE1, showing that
LEU2^{+} and ColE1 are tightly linked.

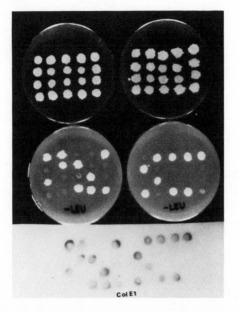

Figure 2. A haploid pYeleu10 transformant was crossed by a leu2⁻
tester strain and the diploid cells were sporulated. After meiosis,
the four individual spores for each of 10 yeast tetrads were dis-
sected and placed on two complete medium plates (2 plates at the
top of the figure). These plates were replicated onto two Leu⁻
plates in order to determine the segregation of the Leu⁺ phonotype
(2 plates at the bottom). The same spores were analyzed by colony
hybridization using ^{32}P-labeled ColE1 DNA as a radioactive probe
(lower panel).

INTEGRATIVE TYPE TRANSFORMATION EVENTS

Transformation by Plasmid DNA Expressed in E. coli

The initial yeast transformation experiments were carried out
using the recombinant plasmids pYeleu10 and pYehis1, which were
isolated from a bank of hybrid plasmids constructed by cloning
randomly sheared yeast DNA into the EcoRI restriction site of the
plasmid ColE1 (9). The plasmids pYeleu10 and pYehis1 were isolated from this collection of plasmids by transforming leuB and
hisB auxotrophs of E. coli. Gene leuB codes for the leucine biosynthetic enzyme β-isopropyl malate dehydrogenase and corresponds
to the leu2 gene in yeast; hisB codes for imidazole glycol phosphate
dehydratase and corresponds to the his3 gene in yeast. For yeast
transformation, stable recipients were developed by recombining
within either the leu2 gene or the his3 gene two stable point

mutations. The resulting double mutants are extremely stable and
revert at a frequency of less than 10^{-10} (none observed to date).
Transformation of these stable auxotrophs with the circular plasmids pYeleu10 and pYehis1 occurs with a frequency of about 10
transformants/µg DNA/10^7 regenerated spheroplasts.

Yeast Transformation With Yeast Genes Not Expressed in E. coli

The his4 gene cluster in yeast (his4A, his4B and his4C), which
codes for three enzymes in the histidine biosynthetic pathway
(phosphoribosyl ATP pyrophosphorylase, phosphoribosyl AMP cyclo-
hydrolase and histidinol dehydrogenase, respectively), has been the
focus of extensive genetic and biochemical analysis. Plasmids
carrying his4 sequences are an obvious choice for further study in
E. coli and yeast. Nonetheless, hybrid plasmid gene banks of yeast
DNA have failed to complement a defect in the hisD gene of E. coli
(G. Fink, unpublished data), the gene corresponding to his4C in
yeast. However, functional his4 sequences have been identified in
such a plasmid bank by transformation in yeast (Hinnen, A., unpublished data). In this experiment, the source of the DNA was a collection of approximately 5000 E. coli clones containing hybrid plasmids
constructed by inserting BamHI fragments of yeast DNA into the Bam
site of the plasmid pBR313 (J. Friesen, personal communication).
The bank strains were arranged into 100 separate pools and plasmids
isolated from each. Each plasmid preparation was then used to
transform a yeast strain carrying a deletion in the his4 region to
His$^+$. Once the pool containing his4 was located, a second round
of plasmid purification and yeast transformation yielded purified
his4 DNA. Yeast strains carrying mutations in the his4A and the
his4C are all transformed to His$^+$ by pYehis4 (P. Farabaugh, unpublished data). Thus, the entire his4 gene cluster is contained on
the plasmid. Transformation by his4 is of the integrative type.

Increased Transformation Frequency With Linear DNA

Linearization of circular plasmids can increase the frequency
of yeast transformation over that obtained with uncut plasmid DNA
(Figure 1). If pYeleu10 is digested with the restriction endonucleases SalI, XhoI or BamHI, the transformation frequency is
increased 5- to 20-fold (3). None of these enzymes cuts within
the essential region of the LEU2$^+$ gene, which spans two RI fragments. Rather remarkably, even an RI digest of pYeleu10 transforms
at least as well as uncut plasmid DNA (Ilgen and Hinnen, unpublished data.)
The increase in transformation frequency by linear DNA is
unexpected. Conventional models predict that a single crossover event is sufficient to integrate a circular plasmid, whereas
a double crossover event is required for linear DNA. However, it
is known that in Bacillus subtilis, linear DNA is as efficient as

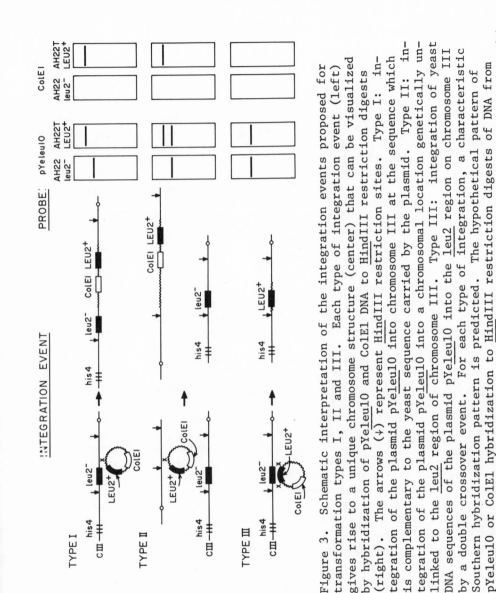

Figure 3. Schematic interpretation of the integration events proposed for transformation types I, II and III. Each type of integration event (left) gives rise to a unique chromosome structure (center) that can be visualized by hybridization of pYeleul0 and ColEl DNA to HindIII restriction digests (right). The arrows (↓) represent HindIII restriction sites. Type I: integration of the plasmid pYeleul0 into chromosome III at the sequence which is complementary to the yeast sequence carried by the plasmid. Type II: integration of the plasmid pYeleul0 into a chromosomal location genetically unlinked to the leu2 region of chromosome III. Type III: integration of yeast DNA sequences of the plasmid pYeleul0 into the leu2 region on chromosome III by a double crossover event. For each type of integration, a characteristic Southern hybridization pattern is predicted. The hypothetical pattern of pYeleul0 or ColEl hybridization to HindIII restriction digests of DNA from each type of transformant is in agreement with the actual patterns obtained (1).

circular DNA in transformation (17). A free end of yeast DNA is important for transformation: no increase in transformation frequency is generated when the E. coli vector portion of the plasmid is cut; cutting must occur within the yeast DNA portion of the plasmid to have any effect on the transformation frequency.

Characterization of Integrative Type Transformants by Genetic Analysis and Southern Filter Hybridization

The Southern filter hybridization procedure, in conjunction with genetic analysis, has revealed three types of pYeleu10 transformants as is shown in Figure 3. Type I and type II transformants are addition transformants in which transformation occurs by a recombination event that adds ColE1 along with leu2$^+$ to the yeast genome. Type III transformants are substitution transformants that have the mutated leu2$^-$ region replaced by the LEU2$^+$ region of the pYeleu10 plasmid

Type I transformants have the entire pYeleu10 plasmid integrated at the leu2 region in chromosome III, resulting in a tight linkage of the leu2$^-$ genes in the resident chromosomes with the incoming LEU2$^+$ genes of the plasmid. The plasmid pYeleu10 has no HindIII restriction endonuclease sites. Thus, in type I transformants, a HindIII restriction fragment corresponding to the leu2 region is replaced by a larger fragment containing pYeleu10 in addition to the resident mutant leu2 region. The duplicated leu2 region in type I transformants is genetically unstable and leads to Leu$^-$ segregants at a frequency of 1%. In type II transformants, the LEU2$^+$ gene segregates independently from the resident leu2$^-$ gene. In type II transformants, a new DNA restriction fragment separable from the restriction fragment carrying the leu2$^-$ gene is generated. Several new locations for the new leu2 sequence have been revealed by Southern analysis. Genetically, the new LEU2$^+$ sequence maps at at least 10 different loci. Type III transformants can be explained by a double crossover event, or by a type I event followed by loss of ColE1 and the resident leu2$^-$ sequence by mitotic segregation. Such an event is genetically and biochemically indistinguishable from a revertant, but no revertants have yet been detected in the strain containing the double mutation at the leu2 locus. Type II and type III transformants display a stable Leu$^+$ phenotype.

Sequence Homology Directs Integration

The close linkage of the transforming sequences to their corresponding sequences in the recipients shows that DNA:DNA homology is important in the integration event. The plasmid pYehis1, which carries the his3 genes of yeast, integrates only into the homologous his3 region on chromosome XV. Similarly,

the his4 plasmid integrates only at the his4 region on chromosome
III. In contrast, only 50% of the pYeleu10 transformants have the
transforming leu2 gene integrated at the leu2 region on chromosome
III. In the remaining pYeleu10 transformants (type II transformants),
the transforming LEU2$^+$ maps elsewhere in the genome. The plasmid
pYeleu10 differs from the other plasmids in that it includes a
segment of yeast DNA that is reiterated at least 20 times in the
yeast genome (18). In type II pYeleu10 transformants, transposition of the LEU2$^+$ sequence to chromosomal sites other than the
leu2 region on chromosome III could result from the presence of
this reiterated sequence on pYeleu10. The integration of the
pYehis4 plasmid only at his4 on chromosome III and of pYehis1 only
at his3 on chromosome XV is probably due to the absence on these
plasmids of the repeated sequence present on pYeleu10.

     The role of sequence homology in the integration process has been
further illustrated by a transformation experiment in which recombination occurs between the bacterial sequences introduced into the
yeast genome by transformation. A strain carrying mutations both
at leu2 and his3 was transformed with pYehis1 to His$^+$ thus adding
ColE1 sequences at the his3 locus on chromosome XV. Subsequently,
an His$^+$ transformant was used in a second transformation experiment
as a host for pYeleu10. Among His$^+$ Leu$^+$ double transformants were
those in which the LEU2$^+$ genes were tightly linked to his3 on chromosome XV (3). Only ColE1 homology within the two plasmids can account
for integration of pYeleu10 at the his3 locus. Similarly, if the
double mutant host were transformed first with pYeleu10 and then with
pYehis1, the HIS3$^+$ region in some double transformants was tightly
linked to LEU2$^+$, ColE1 apparently directing the integration of pYehis1
sequences as well.

                      NONINTEGRATIVE TRANSFORMATION

                 Yeast 2μ Plasmid as a Cloning Vector

     Yeast contains an endogenous plasmid, Saccharomyces cerevisiae
plasmid, or Scp1 (19), that exists as a closed circular molecule
with a circumference of 2 μm, hence referred to as 2μ DNA. Fifty
to 100 copies of the plasmid exist per haploid genome, associated
with neither nuclear nor mitochondrial DNA (20).
     When yeast is transformed with DNA carrying either total 2μ
circular DNA or various restriction fragments derived from it (2,3,
21), the transformation frequency is increased dramatically, to as
many as 20,000 transformants per μg DNA (2). Yeast strains transformed with 2μ circular DNA are genetically unstable, rapidly losing
the transformed phenotype in the absence of selective pressure.
Hinnen, Hicks and Fink (3) have shown by genetic analysis that transforming sequences in 2μ transformants exhibit a nonMendelian pattern
of inheritance indicative of an extrachromosomal location for this

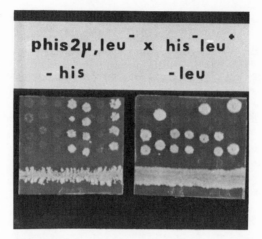

Figure 4. NonMendelian inheritance of the His$^+$ phenotype in a cross of a phis2μR1.4 transformant with an his3$^-$ tester strain. There are two genes segregating in this cross: a LEU2$^+$ gene and a HIS3$^+$ gene. The LEU2$^+$ genes show the normal 2$^+$:2$^-$ segregation of a chromosomal marker. However, the HIS3$^+$ gene segregates either 4$^+$:0$^-$ or 4$^-$:0$^+$, which is indicative of a nonchromosomal location. The non-Mendelian segregation of four His$^+$ spores in three of the tetrads shown is interpreted as a random distribution of the phis2μR1.4 plasmid at meiosis. The absence of His$^+$ spores among the remaining tetrads is due to the loss of phis2μR1.4 prior to meiosis.

DNA. Figure 4 illustrates the genetic behavior of 2μ transformants. In this experiment, the two RI fragments of 2μ circular DNA were cloned into the EcoRI site of pYehis1 to give two new hybrid plasmids (phis2μR1.4 and phis2μR2.6) which were then used to transform an his3 mutant strain to His$^+$. When these transformants were crossed with an his3$^-$ tester strain, the HIS3$^+$ genes segregated either 4$^+$:0$^-$ or $\overline{4^-:0^+}$. A 4$^+$:0$^-$ segregation is assumed to result from distribution of the HIS3$^+$ plasmid to all four meiotic progeny. A segregation pattern of $\overline{4^-:0^+}$ indicates loss of the plasmid prior to meiosis. Recently, Struhl et al. (2) found that transforming 2μ DNA can be recovered as autonomously replicating yeast plasmids at 5 to 10 copies per cell. Interestingly, all transformants carry an integrated copy as well. Moreover, the rare stable transformants, those that maintain the transformed phenotype after several generations of nonselective growth, still contain both integrated and non-integrated copies of the plasmid. It has not been determined whether expression of the transforming sequence is due to the plasmid, the integrated copy, or both. On the basis of a genetic analysis using the KAR mutation (22), Fink suggested that the 2μ circles are nuclear and nonchromosomal.

Autonomously Replicating Cloned Yeast Genes as Vectors for Transformation

Certain hybrid plasmids carrying yeast genes transform yeast at a very high frequency. Struhl et al. (2) have analyzed the transformation event with plasmids of this type carrying the trp1 gene of yeast. The trp1 region was originally isolated by complementation of a trpC mutant in E. coli, and it is of particular interest since trp1 is tightly linked to the centromere of chromosome IV. In yeast, hybrid DNA carrying trp1 transforms at a frequency of 500 to 2000 transformants per µg DNA. The transformation frequency is equally high to Trp$^+$ or to His$^+$, if the HIS3 gene is inserted into this plasmid. Plasmids carrying trp1 seem to behave as minichromosomes in yeast. As with 2µ vectors, they are located outside of the bulk of yeast chromosomal DNA, but in contrast to transformation with 2µ DNA, cells transformed with trp1 do not contain an integrated copy of the plasmid.

Retrieval of Transforming DNA Sequences From Yeast Cells

Yeast DNA present in transformed yeast cells as plasmids can be isolated as closed circular DNA molecules and easily purified away from the bulk of chromosomal DNA (2,19). Recovery of sequences that have become integrated into the yeast genome is more cumbersome but possible if the integrated region includes a marker (such as antibiotic resistance) that can be selected in E. coli. Transformation of E. coli with a restriction digest of yeast DNA then permits the yeast DNA of interest to be isolated along with the selective marker. Subsequently, E. coli can be used as a host for amplification of the associated yeast DNA. In one experiment of this type, selection for tetracycline resistance in E. coli permitted the recovery from transformed yeast cells of an integrated hybrid plasmid carrying a portion of the yeast leu2 region in the tetracycline-resistant plasmid pMB9 (23).

DISCUSSION

The yeast transformation system has greatly expanded experimental approaches to the molecular biology and genetics of yeast. By transformation, specific cloned sequences of yeast DNA can be introduced into the yeast genome, then assayed both biochemically and functionally. Genes that have now been identified by yeast transformation include those previously selected by complementation of E. coli auxotrophic markers (e.g., leu2 and his3) as well as those which fail to complement corresponding E. coli mutations but do function in yeast (his4). Purified hybrid plasmid or phage DNA, plasmid banks, or even ligation mixtures containing yeast and vector

DNA (P. Farabaugh, unpublished data) have all been used as sources
of DNA for yeast transformation.

The transformation frequency is affected by the form of the
DNA (closed circular vs. linear) and by the type of vector used.
In a model system in which yeast strains with stable, double point
mutations at either the leu2 or his3 locus are transformed with
hybrid plasmids carrying the corresponding yeast genes (pYeleu10
and pYehis1), 5 to 20 transformants are obtained per μg closed
circular plasmid DNA. If the plasmid is first digested with a
restriction enzyme that cuts within the yeast DNA portion of the
molecule and outside of the gene, the transformation frequency is
increased 20-fold. If segments of yeast ribosomal DNA or the
yeast endogenous plasmid Scp1 (2μ DNA) are incorporated into vectors,
the transformation frequency is increased 100- to 1000-fold over
that obtained with closed circular plasmid DNA. Similarly, plasmids
carrying certain yeast genes, such as trp1, transform yeast at a high
frequency. The high frequency transformation by trp1 and 2μ plasmids
may result from their capacity to replicate autonomously. However,
in cells transformed with 2μ DNA an integrated copy is also present,
so the reconbination event per se may not be the limiting factor
in transformation.

Foreign DNA vectors have considerable utility in yeast transformation. These sequences are not present in untransformed strains
so their presence in a yeast strain is indicative of a transformation event. The bacterial vector DNA can serve as hybridization
probes in both yeast colony hybridization and in Southern hybridization analysis to signal the integration of transforming DNA into the
genome. Foreign DNA sequences can also be exploited in the genetic
manipulation of local regions of yeast chromosomes. These sequences
provide regions of homology for the insertion or transposition of
other yeast DNA inserted into plasmids with the same foreign DNA
sequence, greatly expanding the potential for structural and
functional analysis of yeast genes. Moreover, transforming DNA
sequences can be retrieved from the yeast genome by selecting in
E. coli for associated sequences such as the antibiotic resistance
markers present on the plasmids.

Transformation will have considerable application in expanding
the molecular biology of yeast. The tailoring of transforming DNA
to promote a high efficiency of transformation now allows yeast
transformation frequencies approaching those obtained in E. coli.
Yeast may thus eventually emerge as an eukaryotic analogue of
E. coli, in which the cloning and possibly the expression of higher
eukaryotic genes can be achieved in a well-characterized genetic
and biochemical background. In addition, yeast has special advantages as a host for cloning, due to its low potential as a biohazard. Saccharomyces cerevisiae is a nonpathogenic laboratory
strain that does not mate or exchange DNA with any bacterial or
fungal pathogen and harbors no known transmissible virus. Until
now, these same features have limited development of the molecular
biology of this organism since there was no yeast system similar

to bacteriophage available for exploitation. Yeast transformation, coupled with the engineering of a yeast phage technology is filling this void in the genetic manipulation of yeast.

Acknowledgments: This work was supported by NIH grants GM15408 and CA23441. All recombinant DNA work was carried out under P2 conditions as approved by the Cornell University Biohazard Committee, the National Science Foundation and the National Institutes of Health. The authors would like to thank Jack Szostak for his help with many of the experiments described here.

REFERENCES

1 Hinnen, A., Hicks, J.B. and Fink, G.R. (1978) Proc. Nat. Acad. Sci. U.S.A. 75, 1929-1933.
2 Struhl, K., Stinchcomb, D.T., Scherer, S. and Davis, R.W. (1978) Proc. Nat. Acad. Sci. U.S.A. (in press).
3 Hinnen, A., Hicks, J.B., Ilgen, C. and Fink, G.R. (1978) in Genetics of Industrial Microorganisms, Amer. Soc. for Microbiology, Washington, D.C. (in press).
4 Grunstein, M. and Hogness, D.S. (1975) Proc. Nat. Acad. Sci. U.S.A. 72, 3961-3965.
5 Beckmann, J.S., Johnson, P.F. and Abelson, J. (1977) Science 196, 205-208.
6 Phillipson, P., Thomas, M., Kramer, R.A. and Davis, R.W. (1978) J. Mol. Biol. (in press).
7 Holland, M., Holland, J. and Jackson, K. (1978) in Methods in Enzymology (Wu, R., ed.), Academic Press, New York, NY (in press).
8 Montgomery, D.L., Hall, B.D., Gillam, S. and Smith, M. (1978) Cell 14, 673-680.
9 Ratzkin, B. and Carbon, J. (1977) Proc. Nat. Acad. Sci. U.S.A. 74, 487-491.
10 Struhl, K., Cameron, J.R. and Davis, R.W. (1976) Proc. Nat. Acad. Sci. U.S.A. 73, 1471-1475.
11 Carbon, J., Clarke, L., Chinault, C., Ratzkin, B. and Waltz, A. (1977) in International Symp. on Pure and Applied Biochemistry of Yeasts (Horecker, B.L. and Breida, M., eds.), Academic Press, New York, NY (in press).
12 Southern, E.H. (1975) J. Mol. Biol. 98, 503-517.
13 Sherman, F., Fink, G.R. and Lawrence, C.W. (1974) Methods in Yeast Genetics, Cold Spring Harbor Laboratory, Cold Spring Harbor, NY.
14 Sherman, F. (1975) in Methods in Cell Biology (Prescott, D.M., ed.), Vol. 11, pp. 189-199, Academic Press, New York, NY.
15 Mortimer, R.K. and Hawthorne, D.C. (1969) in The Yeasts (Rose, A.H. and Harrison, J.G., eds.), Vol. 1, pp. 386-460, Academic Press, New York, NY.
16 Hicks, J. and Fink, G.R. (1977) Nature 269, 265-267.

17   Duncan, C.H., Wilson, G.A. and Young, F.E. (1977) Gene 1,
     153-167.
18   Hicks, J.B., Hinnen, A. and Fink, G.R. (1978) Cold Spring
     Harbor Symp. Quant. Biol. (in press).
19   Cameron, J.R., Phillipson, P. and Davis, R.W. (1977) Nucl.
     Acids Res. 4, 1429-1448.
20   Clark-Walker, G.D. and Miklos, G.L.G. (1974) Eur. J. Biochem.
     41, 359-365.
21   Beggs, J. (1978) Nature 275, 104-108.
22   Conde, J. and Fink, G.R. (1976) Proc. Nat. Acad. Sci. U.S.A.
     73, 3651-3655.
23   Fink, G.R., Hicks, J.B. and Hinnen, A. (1978) in International
     Symp. on Genetic Engineering: Scientific Developments and
     Practical Applications, Elsevier (in press).

THE USE OF SITE-DIRECTED MUTAGENESIS IN REVERSED GENETICS

C. Weissmann, S. Nagata, T. Taniguchi, H. Weber and
F. Meyer

Institut für Molekularbiologie I
Universität Zürich
8093 Zürich, Switzerland

I. INTRODUCTION

In classical genetics, the relationship between genotype and
phenotype is explored by selecting or screening for organisms with
deviant properties and subsequently mapping the cognate lesion in
their genome. This approach has been immensely fruitful in correlating structure-function relationships at the molecular level,
particularly in the case of microorganisms and viruses. It en-
counters limitations when, for example, the role of noncoding,
functionally undefined segments of a genome are to be studied since
we do not know what properties to screen or select for, or when a
lesion leads to an unconditionally lethal mutation. Several years
age we developed a new methodology to deal with such difficulties
in the case of phage Qβ (1-4). In this approach, which we have
called "reversed genetics" (5,6), a mutation is first generated in
a predetermined area of the genome by site-directed mutagenesis
(1) and the effect of the lesion is then studied either in vivo
or in vitro. A similar approach has been applied to the study of
SV40 by Nathans, Berg and their colleagues (7-10). In the first
part of this article, we shall describe the application of reversed
genetics to phage Qβ; in the second part, we shall describe the
extension of site-directed mutagenesis to cloned eukaryotic DNA.

133

II. APPLICATION OF REVERSED GENETICS TO PHAGE Qβ

Some Facts About Phage Qβ and Its Replication

Phage Qβ, a small spherical virus, contains an RNA molecule of
about 4500 nucleotides which serves both as genome and messenger
DNA. As shown in Figure 1, Qβ RNA consists of three translatable
and four nontranslatable (extracistronic) segments. While the
regions immediately preceding the cistrons are involved in the
initiation and regulation of protein synthesis, the function of
the longer untranslated segments at the ends of the genome could
not be determined by classical genetics. It had been suggested
that the precise conservation of these sequences is essential for
the viability of RNA phages because, as far as they were analyzed,
all RNA phages of group I (f2, MS2, R17) had identical terminal
extracistronic regions, even though mutations in coding and inter-
cistronic regions were not uncommon (15).

After penetrating its host, the viral RNA first serves as
messenger RNA. As shown in vitro, ribosomes initially bind almost
exclusively at the initiation site of the coat cistron; binding
at the replicase cistron is thought to occur only after translation
of the coat cistron changes the secondary structure of the RNA, and
the maturation (or A_2) protein is probably only translated from
nascent RNA strands. RNA replication begins after Qβ replicase has
been assembled from the phage coded polypeptide and three host
specified proteins, Tu, Ts and the ribosomal protein S_1.

Purified Qβ replicase, in conjunction with the host factor
HF-1, replicates Qβ RNA in vitro yielding infectious progeny RNA in

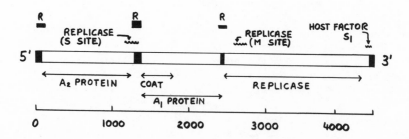

Figure 1. Map of Qβ RNA. Nontranslated areas are black; the cistrons are indicated by double-headed arrows. The ribosome binding
sites are marked R̲, binding sites for Qβ replicase, host factor
and protein S_1 (a component of Qβ replicase) are indicated by wavy
lines (11,27,46).

large excess over the input template; Qβ replicase has low affinity
for the 3' end of Qβ RNA, where RNA synthesis begins, but binds
tightly to two internal sites of Qβ RNA, the S and the M site (16-
18). It is believed that this interaction places the 3' terminus
of the RNA into the initiation site of the polymerase. The product of the first step of synthesis, a single stranded Qβ minus
strand, is noninfectious but serves as an excellent template for
the synthesis of infectious Qβ RNA.

The Technology of Site-Directed Mutagenesis in Qβ RNA

The approach developed in our laboratory is based on the introduction of a mutagenic nucleotide analogue into a predetermined
position of the polynucleotide chain (1). The nucleotide analogue
N^4-hydroxyCTP ($\overline{HO}$CTP) can assume either of two tautomeric forms:
the imino form (a) which can base pair with A, and the amino form
(b) which hydrogen-bonds to G. As a consequence of this tautomeric
equilibrium, $\overline{HO}$CMP can be incorporated into RNA in lieu of either
CMP or UMP (19). As a constituent of an RNA strand, $\overline{HO}$CMP can
direct the incorporation of AMP or GMP with approximately equal
efficiency (1).
The site-specific introduction of the nucleotide analogue is
based on substrate-controlled synthesis (20). Typically, Qβ replicase is incubated with Qβ RNA under conditions of RNA synthesis
but with one (or two) of the standard nucleoside triphosphates
omitted. Elongation stops at the point where the missing triphosphate is required (Figure 2). The replication complex can then be
separated from the unused substrate by Sephadex chromatography and
incubated with a different combination of nucleotides to allow
further limited elongation. This procedure is repeated until the
desired position is reached, whereupon the nucleotide analogue is
incorporated. The minus strand is completed with the four standard nucleoside triphosphates, purified free of plus strands and
used as template for a single round of plus strands synthesis;
extensive replication is avoided to preclude the generation of
aberrant Qβ RNAs (21,22). The resulting RNA preparation consists
of a mixture of wild-type and usually about 20 to 30% mutant RNA;
the presence of a nucleotide substitution is recognized by appropriate oligonucleotide fingerprinting and sequence determination.
In order to test whether or not the mutant RNA is infectious,
spheroplasts are transfected with the RNA mixture, the progeny
virus issuing from single spheroplasts are multiplied and their
RNA is characterized. If the mutated RNA is viable, then a proportion of the phage clones contain the mutant genome; conversely,
if no mutant phage can be detected after screening a large number
of clones, it may be concluded that the mutant RNA has a reduced
infectivity for spheroplasts, or that the burst size is so strongly
reduced that phage detection by a plaque assay is not possible.

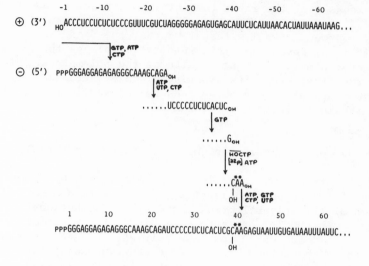

Figure 2. Scheme for the stepwise synthesis of Qβ minus strands with introduction of H̄O̅CMP in position 39 from the 5' terminus.

Mutations in the 3' Extracistronic Region

Qβ RNA with an A→G substitution in position 40 from the 3' terminus. i) Synthesis of the mutated RNA. The procedure used to introduce the nucleotide analogue H̄O̅CMP in position 39 of the minus strand involves five steps and is illustrated in Figure 2. The purified, substituted minus strands were used as template for the synthesis of one round of ^{32}P-labeled plus strands. The product was purified, digested with RNase T_1, and the oligonucleotides were fractionated by two-dimensional polyacrylamide gel electrophoresis. A comparison of the resulting fingerprint with that of wild-type RNA showed that a new large oligonucleotide, designated T1*, had appeared while the amount of oligonucleotide T1 was diminished. The ratio of T1* to T1 was 1:3. Figure 3 shows the positions of these two oligonucleotides in the fingerprint. T1 is derived from positions -63 to -38 at the 3' end of wild-type Qβ RNA (11,23) (Figure 2) and has the sequence

-60   -50   -40

A-A-U-A-A-A-U-U-A-U-C-A-C-A-A-U-U-A-C-U-C-U-U-A-C-Gp   .

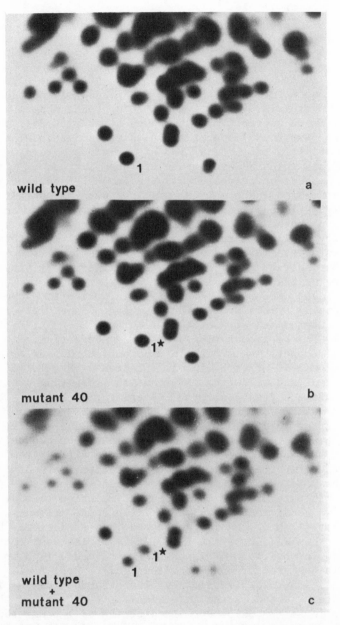

Figure 3. Two-dimensional polyacrylamide gel electrophoresis of the
T_1 oligonucleotides of uniformly ^{32}P-labeled wild-type RNA and mutant
(A_{-40}→G) RNA prepared from cloned phage. (a) Wild-type RNA; (b) mutant (A_{-40}→G) RNA; (c) a mixture of wild-type and mutant (A_{-40}→G)
RNA (3).

The structure of T1* is

   -60   -50   -40

       A-A-U-A-A-A-U-U-A-U-C-A-C-A-A-U-U-A-C-U-C-U-U-Gp   .

Therefore, an A→G transition at position -40 of the wild-type sequence accounts for the appearance of the new oligonucleotide.
Since the synthesis of the minus strands of RNA phages starts at
the penultimate nucleotide of the plus strand (24,25), the 40th
nucleotide from the 3' end of the plus strand is complementary to
the 39th position from the 5' end of the minus strand, i.e., the
position into which N^4-hydroxyCMP had been introduced (Figure 2).
     ii) Isolation of phage carrying an A→G transition in position
-40. One hundred and twenty ng of minus strands substituted with
N^4-hydroxyCMP in position 39 were used as template for one round of
plus strand synthesis. Nine ng of plus strands with a specific
infectivity similar to that of wild-type Qβ RNA were formed.
Spheroplasts were infected with these plus strands and the ^{32}P-labeled
RNA of 18 resultant phage clones was examined by T_1 fingerprinting.
Four preparations showed T1*, diagnostic for the mutant A-40→G, and
no significant amounts of T1; 14 RNAs gave rise to T1 but not to
T1* (Figure 3). This proportion of mutant phage, 22%, reflects
rather accurately the proportion of mutant RNA, 25%, determined
chemically in the preparation used for transfection.
     iii) Competitive growth in vivo between mutant A-40→G and
wild-type phage. E. coli Q13 was infected with a 1:1 mixture of
cloned mutant and wild-type phage and the resulting lysate was
used to infect a fresh culture. After four such cycles of infection, the mutant content was only about 3% and no mutant was
detected after 10 cycles. Propagation of plaque-purified mutant
phage A-40→G in the absence of added wild-type reproducibly resulted in the appearance of wild-type phage after a few cycles
showing that revertants arose at a substantial rate and outgrew
the mutant. We have estimated a reversion rate of 10^{-4} per doubling
and a growth rate of 0.25 of the mutant relative to wild-type under
competitive conditions (3,26). In vitro competition experiments
showed that (A-40→G) RNA is replicated less effectively by Qβ
replicase than wild-type RNA (3). It is of interest that the
nucleotide in position -40 is part of a sequence (-63 to -38) which
binds both host factor and S_1 protein (27). Host factor is required
by Qβ replicase for initiation on plus strands (28); S_1 is a ribosomal protein, which, after infection, is recruited as the α-subunit
of Qβ replicase (29,30). Goelz and Steitz (31) found that the mutant oligonucleotide T1* is bound less efficiently by protein S_1
than its wild-type counterpart, T1, suggesting that the reduced
efficiency of RNA replication could be due to weaker binding of
replicase and/or host factor to the mutated binding site.
     Qβ RNA with a G→A substitution in position 16 from the 3'
terminus. A procedure similar to the one described above was used

to prepare Qβ RNA with a G→A substitution in position 16 from the
3' terminus (1). To determine whether or not this RNA was infectious, spheroplasts were infected with the first generation of
plus strands synthesized on minus strands with HOCMP-substitution
in position -15, as in the previous experiment. One hundred and
twenty clones were analyzed by T1 fingerprinting: all were wild-
type. The plaque formation efficiency of mutant relative to wild-
type RNA was estimated to be less than about 0.03 (22). The
deleterious effect of the (G_{-16}→A) transition on the infectivity
of the RNA does not appear to be due to impaired RNA replication
(2); its cause is not yet known.

   Qβ RNA with substitutions in positions 25 or 29 from the 3'
end (S. Nagata and C. Weissmann, unpublished results). Minus
strands in which positions 24 to 33 were synthesized using HOCTP
rather than CTP were used to direct the formation of plus strands.
After infection of spheroplasts with the plus strand preparation,
30 phage clones were isolated and their RNA analyzed with respect
to the mutagenized region. Three independent mutants were identified: one had a G to A transition in position -25 (i.e., 25
nucleotides from the 3' end), and two had A to G substitutions
in position -29; both had a diminished growth rate when competed
against wild-type Qβ. The specific infectivity of mutant phage
A_{-29}→G was equal to that of wild-type; however that of phage
G_{-25}→A was reduced by about 60%. Equilibrium sedimentation of
phage mutant G_{-25}→A revealed that about one half of the particles
had a diminished buoyant density in CsCl; this fraction of the
phage was noninfectious and contained degraded RNA. These noninfectious particles are reminiscent of the defective phage found
in the case of phage R17 carrying mutations in the A (or maturation
protein) cistron, where, under nonpermissive conditions, exclusively
noninfectious particles of diminished buoyant density, containing
degraded RNA, are formed (32). Possibly, the mutation in position
-25 decreases the affinity of RNA for the A2 protein, so that part
of the particles lack the maturation protein.

   Conclusions about the 3' extracistronic region of Qβ RNA.
We have identified two types of mutations in the 3' extracistronic
region of Qβ RNA (Figure 4), one which is lethal in the sense that
the modified RNA is either no longer infectious or that the burst
size is too small to allow plaque formation, the other which decreases the replication rate of the phage. About 15% of the individual Qβ clones from nonmutagenized populations showed deviations
in the T1 fingerprint pattern; among 25 variants, two had a base
transition in the 3' extracistronic region, one an A→G change in
position -60, the other a C→U change in the region between -39 and
-35. A_{-60}→G showed a diminished competitive growth rate, the other
variant was not tested (33).

   From our results we surmise that at least two functions are
exercised by the 3' extracistronic region: one is in conjunction
with RNA replication, since mutant RNA A_{-40}→G shows a diminished

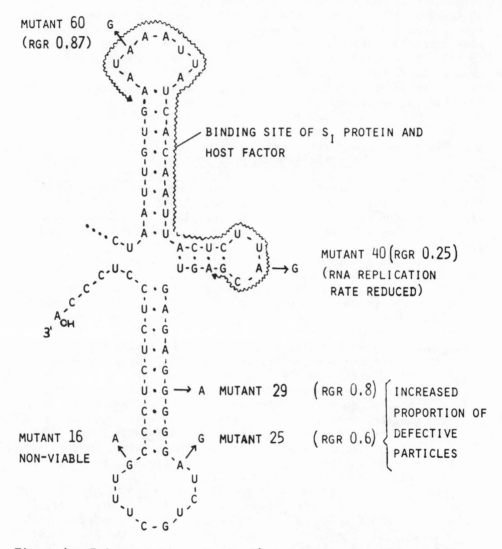

Figure 4. Point mutations in the 3' extracistronic region of Qβ
RNA. Mutations at positions 16, 25, 29 and 40 were generated by
site-directed mutagenesis (1-4; S. Nagata and C. Weissmann, unpublished results); the one at position 60 was found in a wild-
type Qβ population (33). RGR, relative growth rate.

rate of RNA replication and has a lower binding efficiency for
protein S_1, a component of Qβ replicase, while the other is concerned with morphogenesis, since RNA with an $A_{-25} \rightarrow G$ mutation gives
rise to a high proportion of defective particles.

Mutations in the Coat Cistron Initiator Region of Qβ RNA

Synthesis of the RNA. Under initiation conditions, E. coli
ribosomes bind to Qβ RNA almost exclusively at the coat cistron
initiation site (13). To determine to what extent the AUG triplet
is required for 70 S complex formation, we prepared Qβ RNA with
G→A transitions of the third and fourth nucleotides of the coat
cistron, i.e., with modifications in the third positions of the
A-U-G codon and the following nucleotide (Figure 5).
In order to carry out stepwise synthesis in the required
region, we synchronized minus strand synthesis at a ribosome
attached to the coat initiation site. The 70 S Qβ RNA ribosome

Figure 5. Nucleotide sequence around the ribosome binding site of
the coat cistron of wild-type Qβ RNA, and mutants generated by
site-directed mutagenesis at C_3 and C_4. The nucleotide sequence
was established by Hindley and Staples (47) and Weber et al. (16).

complex was used as template for Qβ replicase. Synthesis proceeded up to the position corresponding to the 16th nucleotide of the coat cistron (34). The ribosome was then dislodged by treatment with EDTA and stepwise synthesis was carried out leading to insertion of $\overline{\text{HOCMP}}$ into the positions complementary to the fourth and third nucleotides of the coat cistron (4).

Qβ plus strands were synthesized on the substituted minus strands and replicated in vitro. The ratio of wild-type:mutant $(G_{C_3} \rightarrow A)$:mutant $(G_{C_4} \rightarrow A)$: $(G_{C_3,C_4} \rightarrow A)$ was found to be 1:1.8:1.6:4.5. We have not yet determined whether any of the mutant RNAs are infectious.

The ribosome binding capacity of Qβ RNA with G→A transitions in positions C_3 and/or C_4. A preparation of ^{32}P-labeled Qβ RNA consisting of the species described above was bound to ribosomes under initiation conditions and the 70 S complex was treated with RNase A. The ^{32}P RNA retained by the ribosomes was isolated and analyzed with regard to the C3/C4 region. The relative binding efficiencies of wild-type, mutant $(G_{C_3} \rightarrow A)$, mutant $(G_{C_4} \rightarrow A)$ and mutant $(G_{C_3,C_4} \rightarrow A)$ were estimated to be 1:<0.1:2.8:0.33. Thus, both mutant RNAs lacking the AUG triplet had a reduced ribosome binding capacity. This suggests that the tRNA^{fmet}-AUG interaction contributes substantially to the formation and/or stabilization of the 70 S ribosome complex.

It is striking that ribosomes were bound more efficiently to mutant $(G_{C_4} \rightarrow A)$ RNA than to wild-type RNA, and to mutant $(G_{C_3,C_4} \rightarrow A)$ more efficiently than to mutant $(G_{C_3} \rightarrow A)$ RNA. Perhaps the nucleotides flanking the codon and the anticodon contribute to the stability of the interaction with fMet-tRNA. As shown below, the mutant RNAs $(G_{C_4} \rightarrow A)$ and $(G_{C_3,C_4} \rightarrow A)$ can potentially form an additional A-U base pair as compared to wild-type RNA and mutant $(G_{C_3} \rightarrow A)$ RNA, respectively (4).

$$3' \text{\textbackslash} A-U-A-C-U \text{\textcomma} 5'$$
$$5' -C-A-U-G-G-_{3'}$$
wild-type

$$3' \text{\textbackslash} A-U-A-C-U \text{\textcomma} 5'$$
$$5' -C-A-U-G-A-_{3'}$$
$(G_{C_4} \rightarrow A)$ mutant

$$3' \text{\textbackslash} A-U-A-C-U \text{\textcomma} 5'$$
$$5' -C-A-U-A-G-_{3'}$$
$(G_{C_3} \rightarrow A)$ mutant

$$3' \text{\textbackslash} A-U-A-C-U \text{\textcomma} 5'$$
$$5' -C-A-U-A-A-_{3'}$$
$(G_{C_3,C_4} \rightarrow A)$ mutant

Further Outlooks in Reversed Genetics of Qβ

As mentioned above, site-directed mutagenesis yields a mixture of mutant and wild-type RNA. One heretofore unsolved problem is that of cloning Qβ RNA carrying a lethal mutation in order to study

its biochemical properties. Approaches involving in vitro replication of the RNA were not satisfactory because the extensive
replication required led not only to point mutations but also to
extensive deletions (21,22). In order to clone Qβ RNA, in particular molecules carrying lethal mutations, we have synthesized a
complete DNA copy of Qβ RNA. Poly(A)-elongated Qβ plus and minus
strands were used as templates for reverse transcriptase, the
resulting minus and plus cDNAs were hybridized and elongated with
DNA polymerase to yield a complete double-stranded DNA which was
linked to the plasmid pCRI by the A-T tail method. E. coli transformed by this hybrid grew at a somewhat reduced rate, produced Qβ
phage particles and contained multiple copies of the Qβ DNA plasmid (35). We are using this plasmid to prepare Qβ RNA with lethal
mutations. The mutation can be introduced into the wild-type Qβ
plasmid either by site-directed mutagenesis of the DNA, as described below, or by substituting a segment carrying the lethal
mutation for the corresponding segment of the wild-type Qβ DNA;
such a segment may be obtained by cloning the cDNA copy of a mutated
RNA. In vitro transcription of the Qβ DNA with DNA-dependent RNA
polymerase is expected to yield noninfectious Qβ RNA which can be
studied with regard to its stability to nucleolytic attack, its
ability to enter spheroplasts, to serve as template for RNA replication and protein synthesis, and its capacity to be packaged into
phage particles. Moreover, it will be possible to carry out reversion experiments to prove that the base substitution under consideration is indeed responsible for the lethal effect.

III. APPLICATION OF REVERSED GENETICS TO EUKARYOTIC DNA

     The application of classical genetics to higher organisms is
often tedious: recessive mutations are frequently only recognizable
when homozygous and the genome is so large that the probability of
finding a mutation in a particular DNA region without a potent
selection system is quite low. Moreover, mapping a mutation is
laborious and the identification of mutations in noncoding regions
of the genome is very difficult.
     The advent of hybrid DNA technology is changing this situation
dramatically. It is now possible to integrate a gene of interest
in vitro with its neighboring regions into a vector, clone and
amplify it and reintroduce it into an appropriate host cell to
study its expression, thus allowing the application of reversed
genetics to eukaryotic systems. The modifications introduced into
the DNA may be gross, such as deletions, insertions or trans-
positions of DNA segments by the technique of in vitro recombination,
or point mutations such as the base substitutions generated by site-
directed mutagenesis.

Site-Directed Mutagenesis in DNA by Incorporation
of a Nucleotide Analogue (36)

Introduction of N^4-hydroxyCMP into the positions corresponding
to amino acids 121 and 122 of the rabbit β-globin gene. We have
used the plasmid PβG (37) to determine whether the principle of
site-directed mutagenesis can be applied to DNA. Plasmid PβG,
which contains an almost complete DNA copy of rabbit β-globin mRNA,
has a single EcoRI site within the globin gene region corresponding
to amino acids 121 and 122 (Figure 6). Single-strand nicks were

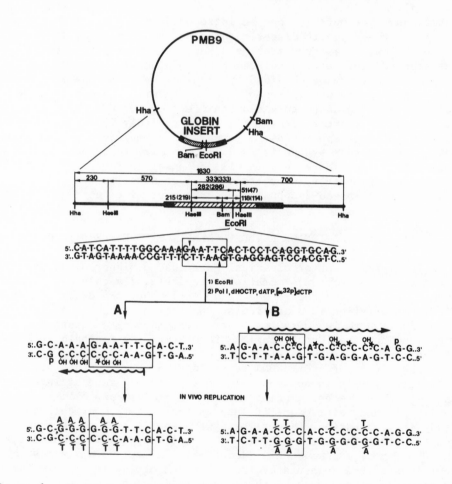

Figure 6.   Partial map of PβG (37) and scheme of procedure used in
site-directed mutagenesis (36).

introduced by partial digestion with EcoRI and the nicked plasmid
was incubated with E. coli DNA polymerase I and dATP, dH̄OCTP and
[α-^{32}P]dCTP, to incorporate the nucleotide analogue into the immediate vicinity of the nick. A control with dTTP instead of
dH̄OCTP was run in parallel.

   Isolation and characterization of EcoRI-resistant PβG DNA.
Four of the nine possible mutations resulting from the replication
of dH̄OCMP-substituted PβG should impart EcoRI resistance to PβG.
To quantitate and isolate such mutants, dH̄OCMP-substituted PβG
as well as the TMP-substituted control and untreated PβG (form I)
were transfected into E. coli HB101 and tetracycline-resistant
(plasmid-containing) colonies were isolated. Samples of plasmid
DNA derived from the three preparations, as well as some of the
original PβG DNA were treated with EcoRI. The residual infectivity
of all control preparations was about 0.07 to 0.08% of the un-
digested samples, while for the DNA derived from the dH̄OCMP-sub-
stituted sample it was 1.9%.

   To ascertain whether the colonies resulting from transfection
with EcoRI-treated DNA were indeed due to EcoRI resistance of
the plasmids, clones from each experiment were cultured and plasmid DNA was isolated. All 24 cloned plasmids derived from the
dH̄OCMP-substituted preparation, but only one of the 24 from the
TMP-substituted sample and 8 of 98 plasmids from the control experiments with PβG were resistant to EcoRI. Most likely, a small fraction of the EcoRI cleaved control DNA had been recircularized in
vivo, restoring EcoRI-sensitive PβG. The average EcoRI resistance
of control DNA was calculated to be 0.005%, and that of the mutagenized sample, 1.9%, a 380-fold higher value.

   Analysis of EcoRI-resistant PβG DNA. Restriction analysis
showed that EcoRI resistance of the plasmids from the control
experiments was in all cases due to extensive deletions which
eliminated the EcoRI site (36,38) probably as a consequence of the
EcoRI cleavage carried out during the selection procedure (7).

   Seven EcoRI-resistant plasmids derived from dH̄OCMP-substituted
preparations were chosen at random and their nucleotide sequence
around the erstwhile EcoRI site was determined. As shown in Table 1,
three plasmids had one, three had two and one had three AT→GC
changes, all located within the mutagenized region (Figure 6).

   Having established that site-directed mutagenesis is in principle feasible, we are now extending the approach to render any
region of the DNA susceptible to nucleotide substitution. Our
immediate intent is to introduce modifications into the putative
ribosome binding site of PβG and to study their effect on protein
initiation. To this purpose, we have prepared a set of cloned,
globin-specific primers, whose 3' termini lie in the region of
interest. These are used as primers on single-stranded circular
DNA for substrate-limited elongation and introduction of the nucleotide analogue.

Table 1

Nucleotide Sequence Around the <u>EcoRI</u> Site of
Wild-Type PβG and <u>EcoRI</u>-Resistant PβG DNA
Generated by Site-Directed Mutagenesis (36).

		119	120	121	122	123	124	125
		gly	lys	glu	phe	thr	pro	gln
wild-type PβG	+	⁵'G–G–C–A–A–A–	G–A–A–T–T–C–	A–C–T–C–C–T–C–A–G
	–	C–C–G–T–T–T–	C–T–T–A–A–G–	T–G–A–G–G–A–G–T–C ₅'

		gly	lys	glu	LEU	thr	pro	gln
PβG 22.1	+	⁵'G–G–C–A–A–A–	G–A–A–C̣–T–C–	A–C–C̣–C–C–T–C–A–G
	–	C–C–G–T–T–T–	C–T–T–G̈–A–G–	T–G–G̈–G–G–A–G–T–C ₅'

		gly	lys	glu	PRO	thr	pro	gln
PβG 22.2	+	⁵'G–G–C–A–A–A–	G–A–A–C̣–C̣–C–	A–C–T–C–C–T–C–A–G
		C–C–G–T–T–T–	C–T–T–G̈–G̈–G–	T–G–A–G–G–A–G–T–C ₅'

		gly	lys	glu	LEU	thr	pro	gln
PβG 22.3	+	⁵'G–G–C–A–A–A–	G–A–A–C̣–T–C–	A–C–T–C–C–T–C–A–G
	–	C–C–G–T–T–T–	C–T–T–G̈–A–G–	T–G–A–G–G–A–G–T–C ₅'

		gly	lys	glu	SER	thr	pro	gln
PβG 22.13	+	⁵'G–G–C–A–A–A–	G–A–A–T–C̣–C–	A–C–T–C–C–T–C–A–G
		C–C–G–T–T–T–	C–T–T–A–G̈–G–	T–G–A–G–G–A–G–T–C ₅'

		gly	lys	glu	PRO	thr	pro	gln
PβG 22.14	+	⁵'G–G–C–A–A–A–	G–A–A–C̣–C̣–C–	A–C–C̣–C–C–T–C–A–G
	–	C–C–G–T–T–T–	C–T–T–G̈–G̈–G–	T–G–G̈–G–G–A–G–T–C ₅'

		gly	lys	GLY	phe	thr	pro	gln
PβG 22.15	+	⁵'G–G–C–A–A–A–	G–G̣–G̣–T–T–C–	A–C–T–C–C–T–C–A–G
	–	C–C–G–T–T–T–	C–C̈–C̈–A–A–G–	T–G–A–G–G–A–G–T–C ₅'

		gly	lys	glu	phe	thr	pro	gln
PβG 12.6	+	⁵'G–G–C–A–A–A–	G–A–G̣–T–T–C–	A–C–T–C–C–T–C–A–G
	–	C–C–G–T–T–T–	C–T–C̈–A–A–G–	T–G–A–G–G–A–G–T–C ₅'

Other Approaches to in vitro Mutagenesis in DNA

Site-directed or region-directed point mutations in DNA have
recently been produced by other approaches. Shortle and Nathans
(39) generated a nick in SV40 DNA using a restriction nuclease,
extended it with DNA polymerase I and mutagenized the exposed
single-stranded regions by treatment with sodium bisulfite (see also
this volume, pp. 73-92). Hutchison, Smith and their colleagues (40)
synthesized a DNA oligonucleotide corresponding to a φX174 minus
strand sequence but containing one nucleotide substitution, hybridized
it to single-stranded φX174 DNA and elongated the DNA to form double-
stranded circles.
Deletions have been generated in specific regions of plasmids
and circular viral DNA by cleaving with a restriction enzyme,
digesting with 5' exonuclease and transforming host cells with the
linear DNA; intracellular ligation recreates circular molecules
which have a deletion at the position of the erstwhile restriction
site (7,8,38). Deletions have been generated at the termini of
the cloned rabbit β-globin gene PβG by excising the gene from the
plasmid by S1 excision, trimming the ends of the DNA segment with
3' exonuclease and S1 nuclease, adding HindIII linkers and recloning the fragments (F. Meyer, H. Heijneker and C. Weissmann,
unpublished results).
Insertions may be introduced at a specific site by cleaving
circular DNA with a restriction enzyme and inserting a DNA fragment using DNA ligase (41), or in the case of cleavage sites with
overhanging 5' ends, by rendering the termini double-stranded with
DNA polymerase and joining them by flush end ligation using T4 DNA
ligase.

IV.   CONCLUSIONS

Currently, reversed genetics are being applied to the study
of viral genomes such as Qβ, SV40 and φX174. The way is open to
explore the expression of cloned yeast genes in yeast (42,43).
Cloned eukaryotic DNA has been linked to SV40 DNA and introduced
into permissive cells where expression was detected, albeit under
control of an SV40 promoter (P. Berg, personal communication).
Expression of cloned eukaryotic DNA coding for 5S RNA and tRNA
has been achieved following injection into Xenopus oocyte nuclei
(45,46) and the elements required for transcription of sea urchin
tRNA genes are being studied by modification of the cloned DNA (46).
Ideally, reversed genetics should be carried out in a homologous
system (i.e., using DNA of the same species as the receptor cell)
if meaningful results are to be obtained, especially when the control of expression is to be studied. Moreover, the conditions
should be as physiological as possible; that is, the gene should
be flanked by extended regions of DNA and be present in the cell

in a low copy number, either as an episome or integrated in the cell DNA. If the corresponding resident genes are not deleted or irreversibly inactivated, the transplanted gene must be appropriately marked to allow identification of the products derived from it. Nonhomologous systems, consisting of components of related systems, may be easier to explore and should be satisfactory for the investigation of certain phenomena, such as the splicing and modification of mRNA precursors, and translation, some of which may, moreover, be studied in vitro.

REFERENCES

1   Flavell, R.A., Sabo, D.L., Bandle, E.F. and Weissmann, C.
    (1974) J. Mol. Biol. 89, 255–272.
2   Flavell, R.A., Sabo, D.L.O., Bandle, E.F. and Weissmann, C.
    (1975) Proc. Nat. Acad. Sci. U.S.A. 72, 367–371.
3   Domingo, E., Flavell, R.A. and Weissmann, C. (1976) Gene,
    1, 3–25.
4   Taniguchi, T. and Weissmann, C. (1978) J. Mol. Biol. 118,
    533–565.
5   Weissmann, C. (1978) TIBS 3, N109–N111.
6   Weissmann, C., Weber, H., Taniguchi, T., Müller, W. and Meyer,
    F. (1978) in Genetic Engineering (Boyer, H.W. and Nicosia, S.,
    eds.), pp. 65–76, Elsevier/North-Holland Biomedical Press,
    Amsterdam.
7   Lai, C.-J. and Nathans, D. (1974) J. Mol. Biol. 89, 179–193.
8   Carbon, J., Shenk, T. and Berg, P. (1975) Proc. Nat. Acad.
    Sci. U.S.A. 72, 1392–1396.
9   Rundell, K., Collins, J.K., Tegtmeyer, P., Ozer, H.L., Lai,
    C.-J. and Nathans, D. (1977) J. Virol. 21, 636–646.
10 Cole, C.N., Landers, T., Goff, S.P., Manteuil-Brutlag, S.
    and Berg, P. (1977) J. Virol. 24, 277–294.
11 Weissmann, C., Billeter, M.A., Goodman, H.M., Hindley, J.
    and Weber, H. (1973) Annu. Rev. Biochem. 42, 303–328.
12 Weissmann, C. (1974) FEBS Lett. 40, S10–S18.
13 RNA Phages (1975) (Zinder, N.D., ed.) Cold Spring Harbor
    Laboratory, Cold Spring Harbor, NY.
14 Blumenthal, T. and Carmichael, G.G. (1979) Annu. Rev. Biochem.
    48 (in press).
15 Min Jou, W., Haegeman, G., Ysebaert, M. and Fiers, W. (1972)
    Nature 237, 82–88.
16 Weber, H., Billeter, M., Kahane, S., Hindley, J., Porter, A.
    and Weissmann, C. (1972) Nature New Biol. 237, 166–170.
17 Meyer, F., Weber, H. and Weissmann, C. (1976) Experientia
    32, 804.
18 Meyer, F. (1978) Dissertation, University of Zürich, Switzerland.
19 Budowsky, E.I. (1976) Progr. Nucl. Acid Res. Mol. Biol. 16,
    125–188.

20   Bandle, E. and Weissmann, C. (1972) Experientia 28, 743–744.
21   Mills, D.R., Peterson, R.L. and Spiegelman, S. (1967) Proc.
      Nat. Acad. Sci. U.S.A. 58, 217–224.
22   Sabo, D.L., Domingo, E., Bandle, E.F., Flavell, R.A. and
      Weissmann, C. (1977) J. Mol. Biol. 112, 235–252.
23   Goodman, H.M., Billeter, M.A., Hindley, J. and Weissmann, C.
      (1970) Proc. Nat. Acad. Sci. U.S.A. 67, 921–928.
24   Rensing, U. and August, J.T. (1969) Nature 224, 853–856.
25   Weber, H. and Weissmann, C. (1970) J. Mol. Biol. 51, 215–224.
26   Batschelet, E., Domingo, E. and Weissmann, C. (1976) Gene
      1, 27–32.
27   Senear, A.W. and Steitz, J.A. (1976) J. Biol. Chem. 251,
      1902–1912.
28   Franze de Fernandez, M.T., Hayward, W.S. and August, J.T.
      (1972) J. Biol. Chem. 247, 824–831.
29   Kamen, R.I., (1970) Nature 228, 527–533.
30   Kondo, M., Gallerani, R. and Weissmann, C. (1970) Nature
      228, 525–527.
31   Goelz, S. and Steitz, J.A. (1977) J. Biol. Chem. 252, 5177–
      5179.
32   Argetsinger, J.E. and Gussin, G.N. (1966) J. Mol. Biol. 21,
      421–434.
33   Domingo, E., Sabo, D., Taniguchi, T. and Weissmann, C. (1978)
      Cell 13, 735–744.
34   Kolakofsky, D., Billeter, M.A., Weber, H. and Weissmann, C.
      (1973) J. Mol. Biol. 76, 271–284.
35   Taniguchi, T., Palmieri, M. and Weissmann, C. (1978) Nature
      274, 223–228.
36   Müller, W., Weber, H., Meyer, F. and Weissmann, C. (1978)
      J. Mol. Biol. 124, 343–358.
37   Efstratiadis, A., Kafatos, F.C. and Maniatis, T. (1977)
      Cell 10, 571–585.
38   Meyer, F., Müller, W., Palmieri, M. and Weber, H. (1978)
      Experientia 34, 948.
39   Shortle, D. and Nathans, D. (1978) Proc. Nat. Acad. Sci.
      U.S.A. 75, 2170–2174.
40   Hutchison, C.A. III, Phillips, S., Edgell, M.H., Gillam, S.,
      Jahnke, P. and Smith, M. (1978) J. Biol. Chem. 253, 6551–6560.
41   Rodriguez, R.L., Tait, R., Shine, J., Bolivar, F., Heyneker,
      H., Betlach, M. and Boyer, H.W. (1977) in Molecular Cloning
      of Recombinant DNA (Scott, W.A. and Werner, R., eds.), pp. 73–
      85, Academic Press, New York, NY.
42   Hinnen, A., Hicks, J.B. and Fink, G.R. (1978) Proc. Nat. Acad.
      Sci. U.S.A. 75, 1929–1933.
43   Beggs, J.D. (1978) Nature 275, 104–109.
44   Brown, D.D. and Gurdon, J.B. (1977) Proc. Nat. Acad. Sci.
      U.S.A. 74, 2064–2068.
45   Kressmann, A., Clarkson, S.G., Pirrotta, V. and Birnstiel,
      M.L. (1978) Proc. Nat. Acad. Sci. U.S.A. 75, 1176–1180.

46   Hindley, J., Staples, D.H., Billeter, M.A. and Weissmann, C.
      (1970) Proc. Nat. Acad. Sci. U.S.A. 67, 1180–1187.
47   Hindley, J. and Staples, D.H. (1969) Nature 224, 964–967.

AGROBACTERIUM TUMOR INDUCING PLASMIDS:

POTENTIAL VECTORS FOR THE GENETIC ENGINEERING OF PLANTS

P.J.J. Hooykaas, R.A. Schilperoort and A. Rörsch

Department of Biochemistry, Leiden State University

Leiden, The Netherlands

1. INTRODUCTION

By now, it is well established that many Agrobacterium species transfer a small piece of bacterial DNA to plant cells. Although the accompanying phenomenon, the induction of tumors, is in itself a very interesting process which certainly deserves further study, it is currently receiving widespread attention because the bacterial plasmids involved may be used to transfer foreign genes into plants.

Although nothing is known about the stability or the expression of foreign genes in plant cells, and very little is known about the efficiency of genetic transfer mediated by Agrobacterium plasmids, the potentials of the system are clear. The bacterial plasmids can be engineered in vitro and in vivo, reintroduced by transformation into bacteria, and moved around among various Rhizobiaceae, including the nitrogen fixing ones. Here, we will review the relevant properties of the Agrobacterium and related plasmids and their interspecies transfer, with special attention to their use in genetic engineering.

2. THE PHYSIOLOGY OF TUMOR INDUCTION

More than 70 years ago, Smith and Townsend (1) found that the plant disease crown gall is caused by a bacterium now called Agrobacterium tumefaciens. Since then, the other plant diseases, cane gall and hairy root, have been shown to be caused by species of the genus Agrobacterium, A. rubi and A. rhizogenes, respectively. All of these diseases are characterized by host cell proliferation (tumor or gall formation) and have been studied thoroughly because

151

of their relevance to cancer and their economic importance. Crown
gall has been reported to be a widespread disease in the fruit crops
and grapevines of Eastern Europe and Australia (2,3,4).

In nature, crown gall is commonly induced on plants at or just
below the soil surface at the root crown, while cane gall is mainly
aerial. The hairy root disease is characterized by very small tumors
from which copious roots arise. In the laboratory, all three diseases can be induced on healthy plants by infecting them with the
bacteria. The plants are wounded before inoculation with the bacteria. Tumors are formed at the wound sites, usually the stem or
the leaves of the plant, but other parts are also sensitive. Most
tumorigenic Agrobacteria are able to induce tumors on a wide range
of plant species, probably in all dicotyledonous plants (5,6). The
cane gall bacteria were isolated from aerial galls on plants of the
genus Rubus (e.g., blackberry) and therefore were named A. rubi.
It is most likely that these bacteria also have this wide host
range. Tumors induced by this species do not seem to be different
from those induced by A. tumefaciens. Among Agrobacteria isolated
from grapevines, strains were found with a limited host range: a)
strains which only form tumors on grapevine (7,8); b) a strain
(NCPPB 1771) which induces tumors on sunflower, tobacco and tomato,
for example, but not on Kalanchoë, which is a good host for other
Agrobacteria.

Besides host range, virulence must be considered as a distinct
genetic property of Agrobacterium strains. Avirulent strains do
occur in nature and, as we shall describe in more detail later, the
opposite conversion, from a virulent to a nonvirulent strain, can
be brought about experimentally.

Direct application or infiltration of Agrobacteria into plant
stems or leaves only rarely leads to tumorigenesis, while tumors
are formed readily when the bacteria are introduced by wounding (9).
Wounding probably is necessary: a) to supply sites at which the bacteria may attach, b) to enable the bacteria to enter the plant
(Agrobacteria lack invasiveness), c) to supply a medium (of plant
metabolites) that supports bacterial growth and, d) to provide
"conditioned" plant cells, i.e., cells susceptible to tumor induction by A. tumefaciens.

It is suggested that there is a particular cellular stage
between quiescence and division during which the plant cell can
respond to the tumorigenic stimulus. Only during this sensitive
stage, which is finite in duration, can a cell be diverted from its
morphogenetic path to form a tumor. Experiments with antimetabolites
to see if tumor formation is inhibited (10) indicate that the DNA
synthesis period "S" of the mitotic cycle may be the stage during
which the cells are susceptible to tumor induction.

For tumor induction, tumorigenic bacteria must attach to
sensitive plant cells (11,12; Figure 1). Attachment occurs at a
limited number of sites on the plant cell surface. Avirulent Agrobacteria are able to compete with virulent bacteria for the attachment sites and can inhibit tumorigenesis in this way. Some avirulent

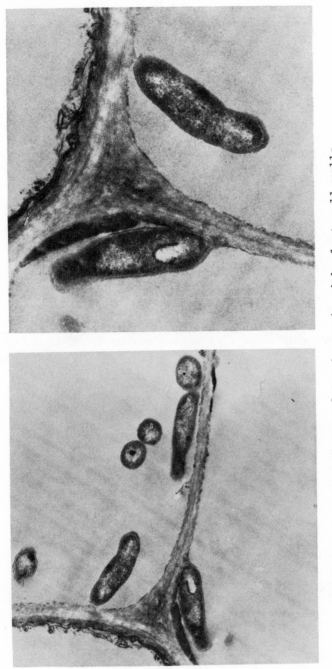

Figure 1. Association of Agrobacteria with plant cell walls.

Agrobacteria, however, lack the ability to compete (11,13; see also
section 7, below). When wound sites are preincubated with cell wall
preparations of virulent or competing bacteria, or with the lipopolysaccharide (LPS) fraction isolated from such preparations, tumor
induction is severely inhibited. These data suggest that the bacteria bind to the plant cells at least partly through LPS (13).
Agrobacteria do not penetrate plant cells but probably transform by
transferring genetic material to them (see sections 6 and 9, below).

    After plant cells have been transformed by Agrobacterium, they
multiply and give rise to tumors even in the absence of the inciting
bacteria. Crown gall tumor tissue is characterized by its ability
to grow autonomously even in the absence of plant growth factors.
In most plants, tumors induced by different Agrobacteria have the
same morphology. However, in Kalanchoë daigremontiana the tumor
morphology is dependent on the inciting bacteria (14,15). Some
Agrobacteria induce "rough" tumors, which are characterized by a
rough surface surrounded by many roots (Figure 2a), while others
induce "smooth" tumors having a smooth surface and a few roots
only at the bottom of the tumor (Figure 2b). The latter tumors
can give rise to leaflike structures (teratomata). While plants
can be regenerated fairly easily from smooth-type tumors (e.g.,
tobacco tissue in vitro) it is very difficult to do so from rough
tumors. On Kalanchoë, A. rhizogenes shows root growth almost entirely at wound sites (Figure 2).

    Unusual amino acid derivatives may be found in A. tumefaciens
tumors (16). Depending on which bacteria induced the tumor (rough or
smooth), they either contain: rough - octopine (18), octopinic
acid (19), lysopine (20,21) and histopine (22) (all amino acid
condensates to pyruvate, Figure 3) or none of these compounds;
smooth - nopaline (23) and ornaline (24,25; Figure 3) or neither
of these (Table 1; 14,16,17). Since tumors are known that do not
contain any of these unusual amino acids, they are considered to be
tumor-specific compounds that are not essential for the tumorous
state. For example, in tumors induced by A. rhizogenes, neither
octopine nor nopaline were detected (8, 14).

                          3. TAXONOMY

    Agrobacteria are common soil inhabitants. They are gram-
negative rods, motile with one or more peritrichous flagella, with
an optimum growth temperature range from 25 to 30°C. Together with
Rhizobia they are included in the family Rhizobiaceae (26). The
following species were originally recognized: Agrobacterium radiobacter, A. tumefaciens, A. rhizogenes, A. rubi, Rhizobium meliloti,
R. trifolii, R. phaseoli, R. leguminosarum, R. lupini, R. japonicum.
The species were characterized as such primarily by their phyto-
pathogenicity and their symbiotic properties.

    Numerous taxonomic studies and studies on DNA composition
have shown that Agrobacteria and the fast growing Rhizobia are

Figure 2. A comparison between different types of tumors induced on Kalanchoë diagremontiana.
A) Rough tumor. B) Smooth tumor. C) Hairy root tumor.

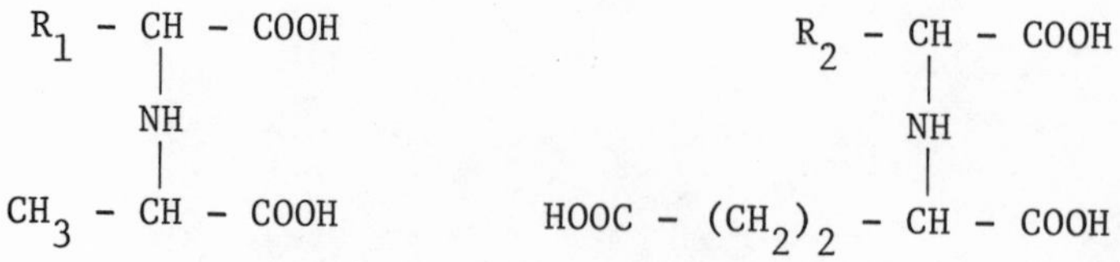

R_1 – CH – COOH R_2 – CH – COOH
       | |
     NH   NH
     | |
CH_3 – CH – COOH   HOOC – $(CH_2)_2$ – CH – COOH

	Name	Catabolized into
R_1 – :		
NH_2 – C – NH – $(CH_2)_3$ – ($\parallel$ NH)	Octopine	Pyruvate + Arginine
NH_2 – $(CH_2)_4$ –	Lysopine	Pyruvate + Lysine
NH_2 – $(CH_2)_3$ –	Octopinic Acid (Ornopine)	Pyruvate + Ornithine
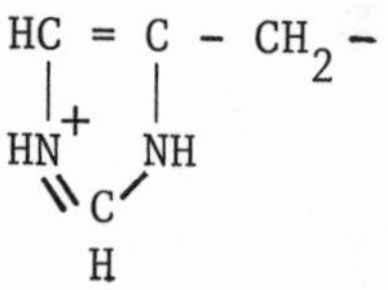 HC = C – CH_2 – (HN$^+$ NH, C, H)	Histopine	Pyruvate + Histidine
R_2 – :		
NH_2 – C – NH – $(CH_2)_3$ – ($\parallel$ NH)	Nopaline	α–Ketoglutarate + Arginine
NH_2 – $(CH_2)_3$ –	Nopalinic acid (Ornaline)	α–Ketoglutarate + Ornithine

Figure 3. Unusual amino acids found in crown gall tumors.

closely related, but that the slow growing Rhizobia (R. lupini and R. japonicum) are more distant relatives (27-32). The relatedness between Agrobacteria and Rhizobia has also been demonstrated by studies showing that certain Rhizobia were able to induce tumors on plants after treatment with UV-irradiation (33). Experiments have also shown that Agrobacteria can be assigned to two clusters or biotypes. In both clusters, nontumorigenic, tumorigenic and

Table 1

Properties of Agrobacterium Strains

Bacterial strain	Biotype	Tumor morphology	Oct/nop[a] synthesis in tumor	Oct/nop catabolism by bacterium
B6, A6, 15955, B2A	1	rough	oct	oct
C58, K14, T37	1	smooth	nop	nop
396, 398, 925, 542	1	rough	-	-
1651, K27, 2303	2	smooth	nop	nop[b]
EU6	2	smooth	-	nop[b]
TR101, 1855	2			
	(A. rhizogenes)	roots	-	-[b]
AG19, AG60, AG67	1	AV[c]		oct
S1005, 2406, 0363	1	AV[c]		-

[a] Oct, octopine; nop, nopaline.

[b] Biotype 2 strains do degrade octopine, but slowly (36). This trait is not determined by the Ti-plasmid (37).

[c] AV, avirulent.

rhizogenic strains are represented. Bacteria belonging to a third biotype, an intermediate between biotype 1 and 2, have also been found (31). Genes on transmissible plasmids have been shown to be responsible for the phytopathogenic properties of the bacteria (see section 6, below) and these properties are therefore less important taxonomically. The fast growing Rhizobia fell into two clusters, one comprised of R. meliloti, and the other of R. trifolii, R. leguminosarum and R. phaseoli. Studies with phage have confirmed these data: Phage isolated from R. meliloti generally do not lyse other Rhizobia or Agrobacteria; phage isolated from R. leguminosarum do lyse strains of R. trifolii and R. phaseoli, but not R. meliloti or Agrobacteria; phage isolated from biotype 1 Agrobacteria do not lyse Agrobacteria of other biotypes or Rhizobia. A limited number of properties for each of these five clusters are compared in Table 2.

4. GENE TRANSFER AMONG RHIZOBIACEAE

Genetic methods developed for E. coli have been found not to be immediately applicable to the bacteria of the Rhizobiaceae. However, a number of techniques can be applied after some slight modification. These will be briefly described in this section.

Mutagenesis

Mutagenesis of Agrobacterium with N–methyl–N'–nitroso–N–nitrosoguanidine (MNNG) has been described by several investigators (38–40). When bacteria, suspended in citrate buffer, are incubated in the presence of MNNG (200 µg/ml) for 3 hr, survival is approximately 3%. A similar method has been described for R. leguminosarum (41). Ethyl methanesulfonate (EMS) mutagenesis is possible for Agrobacterium (42) by suspending the bacteria in a 0.2 M Tris-HCl solution containing K_2HPO_4, 10.5 g/l; KH_2PO_4, 4.5 g/l; $(NH_4)_2SO_4$, 1.0 g/l; pH 7.5, and incubating them for 2 hr with EMS at a final concentration of 0.06 M. Mutagenesis by ultraviolet irradiation (UV) has also been performed by suspending bacteria in distilled water and irradiating them with 60 J/m^2 to approximately 0.1% survival (8).

Replica Plating

Mutants can be detected by replica plating on selective media. For Rhizobia, minimal media should be supplemented with sodium succinate to reduce the amount of exopolysaccharides produced by Rhizobia when grown on minimal media containing glucose or mannitol

Table 2

Comparison of Agrobacteria and Fast Growing Rhizobia

	Agrobacterium biotype			Rhizobium meliloti	Rhizobium leguminosarum[a]
	1	2	3		
3 ketolactose production (30,31)	+	−	−	−	−
Growth on erythritol as carbon source (30,31)	−	+	−	+	+
Growth on 2% NaCl (30, 31)	+	−	+	+	−
Growth at 37°C (30, 31)	+	−	−	+	−
Nile blue reduction (33)	+	+	NT[b]	−	−

[a] R. trifolii and R. phaseoli are included in this group.

[b] NT, not tested.

(41). This increases the number of colonies that can be handled per plate.

Enrichment Procedure

An enrichment procedure for Agrobacterium has been described (39). Bacteria, grown in medium in which selected mutants cannot grow, are treated with carbenicillin at a concentration of 500 µg/ml for 4 hr in the presence of 100 µg/ml lysozyme. After 4 hr, the viable count drops to approximately 0.2% of the value at zero time. This treatment increases the yield of mutants by a factor of around 10^4 after two cycles.

Transformation

Agrobacterium has been transformed with plasmid DNA by a freeze-thaw procedure (43). With plasmid RP4 a maximum frequency of 3.5 x 10^{-7} transformants per total recipient population has been obtained; with Ti-plasmids, the maximum transformation efficiency was 4.5 x 10^{-8}. Transformation of R. trifolii with RP4 plasmid DNA has been described also (44).

Transduction

Although many phage, including temperate phage, have been isolated from Agrobacterium, transduction has not been performed. For R. meliloti, however, both generalized (45) and specialized transduction (46) have been shown.

Conjugation

R-plasmids of incompatibility group P (Inc P) (e.g., RP4, R702, R751.pMG1#2) have been introduced into all of the Agrobacterium and Rhizobium species (8,41,47-49). The resistance markers, except for carbenicillin/ampicillin resistance, are expressed in all strains harboring the plasmids. The plasmids are transferable between Agrobacterium and Rhizobium species at high frequency -- generally 1 to 80% of the recipients receive the R-plasmid provided conjugation has been performed on a membrane filter placed on a solid medium. Agrobacterium strains carrying RP4 are sensitive to phage GU5, but Rhizobia carrying the same plasmid are resistant. A plasmid of incompatibility group W (Inc W) has been shown to be transferable to Agrobacterium (49). Plasmids of other incompatibility groups were neither transferable to Agrobacteria nor to Rhizobia (48,49). An R-plasmid of unknown incompatibility naturally occurring in R. japonicum is also transferable to Agrobacterium (50).

Chromosome Mobilization

This technique has been described extensively for R. meliloti
and R. leguminosarum, and maps have been constructed for R. meliloti
with R68.45 (51) and RP4 (52). For R. leguminosarum, R68.45 was
active but RP4 was not (53,54). With R68.45, chromosomal genes have
been transferred even between different species (55,56). When
chromosomal genes were transferred from R. leguminosarum to
R. trifolii or R. phaseoli or vice versa, stable haploid recombinants
were obtained. However, when chromosomal genes were transferred
between R. leguminosarum and R. meliloti, haploidy did not result,
although bacteria carrying R-prime plasmids were obtained. These
findings once again show the very close relationship between R.
leguminosarum, R. trifolii and R. phaseoli, and the great distance
between these bacteria and R. meliloti (see section 3, above).
Chromosome mobilization for Agrobacterium has not yet been published,
although chromosomal genes could be mobilized with R-plasmids RP4,
R702 and R68.45 in strain C58 (8).

5. NATURALLY OCCURRING PLASMIDS IN THE RHIZOBIACEAE

Plasmids have been detected in almost all bacterial strains of
the Rhizobiaceae family whether they can or cannot induce tumors or
nodules. Most strains carry one or more large plasmids ranging in
size from about 100 to >200 megadaltons (57-63). New procedures
have been developed for the detection, isolation and characterization
of these large plasmids (57,61,62,64,65). Cleared lysate procedures
(66) have been successful in the detection and isolation of plasmids
of low molecular weight in Rhizobium and Agrobacterium (67) but not
for those of high molecular weight. Large covalently closed circular
(CCC) molecules have been detected in Agrobacterium and Rhizobium
by velocity sedimentation of their DNA only after the bacteria had
been lysed by strongly polar detergents (e.g., sodium dodecylsulfate
or sarcosinate), which allow large plasmid DNA to dissociate from
membrane complexes (57,65). The plasmid nature of isolated DNA has
been demonstrated by reassociation kinetic analysis (61,65) and
molecular weights have been estimated by the same technique as it
allows an approximation of the molecular homogeneity of the DNA.
Recently, procedures have been developed for the isolation of large
CCC molecules on a preparative scale (62,64). These methods involve neutralization of the sheared lysate immediately after alkaline
denaturation to prevent nicking and irreversible denaturation of the
plasmid DNA, while linear chromosome DNA remains denatured. The
chromosomal single-stranded DNA is removed by phenol treatment in
the presence of 3% NaCl.
Both large and small plasmids may also be detected by applying
crude DNA extracts of small bacterial cultures directly to agarose

gels for electrophoresis (63). This method is ideal for rapid screening of many strains for plasmids, although plasmids larger than 250 megadaltons may be overlooked.

It is noteworthy that plasmids smaller than 90 megadaltons have only rarely been found in Rhizobia and that very small plasmids suitable as cloning vehicles for genetic engineering (smaller than 10 megadaltons) have not been detected at all.

6. ROLE OF PLASMID GENES IN TUMORIGENESIS

Some Agrobacterium strains lose the ability to induce tumors when grown at 37°C, a temperature higher than the normal growth temperature (29°C) (70). This fact suggested that, at least for these strains, genes involved in tumorigenesis are extrachromosomally located. In agreement with this hypothesis was the finding that virulent Agrobacteria were able to transfer their virulence to avirulent Agrobacteria in mixed infections on plants (71,72,83). When avirulent recipient bacteria are reisolated from tumors induced by a virulent strain in the presence of these avirulent recipients, up to 40% of the recipients become virulent. Moreover, A. rhizogenes bacteria are able to transfer the ability to induce root proliferation in mixed infections with avirulent Agrobacteria (73), also implying that these genes are at least partly extrachromosomal.

Indeed it has been found that A. tumefaciens strains that had lost virulence upon heat treatment, had also lost one of their large plasmids, while avirulent Agrobacteria, which had gained virulence in a mixed infection on the plant, had acquired a large plasmid (37,68,69,74). Such plasmids have, therefore, been called tumor-inducing or Ti-plasmids and range in molecular weight from about 90 to 160 x 10^6 (75).

A very important recent finding indicates that during tumor induction, plasmid DNA is transferred into the plant cell (76-78) where it is probably transcribed (77,78). The region of the plasmid detected in plant tumor tissue is called the T-DNA region. It is doubtful, however, that this part of the plasmid is essential for tumor induction (see section 9).

For one Rhizobium strain at least, genes involved in nodulation seem to be extrachromosonally located, since this strain was no longer able to induce root nodules after loss of one of its large plasmids (62). Therefore, it may be a general phenomenon that bacterial genes coding for the ability to induce plant cell proliferation are located on plasmids of large size.

7. FURTHER CHARACTERIZATION OF Ti-PLASMIDS

When compared with their pathogenic parents, Ti-plasmid-lacking avirulent derivatives differ in a number of properties

(Table 3). Strains cured of their Ti-plasmid are resistant to
agrocin 84, and agrocin 84-resistant mutants have been found to be
cured of their Ti-plasmid (79,80). One method for the biological
control of crown gall consists of inoculating susceptible plants
with strain Kerr 84, which produces this bacteriocin (81). It has
been found that no plaques are formed by phage AP1 on strains
carrying a Ti-plasmid, although they are killed. This phenomenon
has been called AP1 exclusion (15,74,82). Furthermore, Ti-plasmid-
cured strains lose their ability to degrade octopine or nopaline
(37,74,84).

Since transfer of Ti-plasmids has become possible ex planta,
plasmid markers can be studied more extensively. For this, media
selective for exconjugants were needed. Exconjugants from in planta
crosses were chosen as pathogenic recipients. Media selective for
bacteria that can utilize either octopine or nopaline as a carbon
and nitrogen source had been developed. However, growth in these
media was slow. Compounds present in the agar enabled Agrobacteria
to grow in the absence of an added nitrogen source, making it impossible to select for strains using a specific nitrogen source.
Washing the agar did not always yield satisfactory results. Therefore, a new selective medium was developed with either octopine or
nopaline present as a nitrogen source, a pH indicator such as bromo-
thymoblue, and a minimal amount of phosphate buffer. Ti-plasmid-
harboring strains form yellow colonies on this medium, while Ti-
plasmid-lacking strains remain translucent (47). With this selective medium, it was found that Inc P plasmids mobilize Ti-plasmids
(15,47,82,84,85).

Table 3

Comparison of Strains With and Without Ti-Plasmids

	Agrobacterium				
	C58	C58 cured	Ach5	Ach5 cured	C58 (pTiAch5)
Ti-plasmid	+	−	+	−	+
Oncogenicity	+	−	+	−	+
Tumor morphology	smooth		rough		rough
Unusual guanidines in tumors	nopaline		octopine		octopine
Octopine catabolism	−	−	+	−	+
Nopaline catabolism	+	−	−	−	−
Agrocin 84 sensitivity	+	−	−	−	−
Phage AP1 sensitivity	+	+	+	+	+
Phage AP1 exclusion	+	−	+	−	+

Ex planta transfer of Ti-plasmids on rich medium has never been observed in the absence of a mobilizing R-plasmid. However, in the presence of RP4, R702, R68.45 or R751.pMG1≠2 transfer has been observed although recipients receiving a Ti-plasmid and an R-plasmid behave as if the Ti-plasmid has acquired affinity for the R-plasmid. When such recipients are used as donors in further crosses, 20 to 100% of the recipients receiving the R-plasmid also receive the Ti-plasmid, while the frequency of cotransfer is less than 10^{-6} for the original donor (86).

Affinity appeared to be based on the following transpositional and recombinational events:

1) In some recipients, cointegrate plasmids consisting of an R-plasmid and a Ti-plasmid have been observed (86-88). Frequently, such cointegrates do not consist of both complete original plasmids but one or the other or both may be partially deleted. These cointegrates are fairly stable and can be used to introduce Ti-plasmids into new hosts.

2) Other recipients have been found to carry Ti-plasmids with an inserted transposon originating from the R-plasmid. From these, Ti-plasmids have been obtained carrying Tnl from RP4, and others carrying a transposon of 10.3 megadaltons from R702, coding for streptomycin-spectinomycin resistance. In fact, new transposons may be discovered by the use of R-plasmids with the transposons as vehicles, perhaps mobilizing other plasmids (86).

It was shown recently that Ti-plasmids themselves are conjugative but only when crosses are performed on minimal media containing octopine, lysopine or octopinic acid for octopine Ti-plasmids and nopaline for nopaline Ti-plasmids (47,89,90). Certain amino acids (methionine, cysteine and cystine) have been found to inhibit transfer, while on rich media, even in the presence of octopine, transfer has never been observed (47). For some strains, the addition of $MnSO_4$ to minimal medium, together with octopine, is essential for transfer to occur (87).

Since transfer takes place on minimal medium (without octopine), when donors pregrown in media containing octopine are used it would seem that octopine is an inducer of the transfer genes.(91). Induction is very specific: neither arginine nor pyruvate nor the combination of the two are able to induce transfer. Even octopine analogues such as nor-octopine and (desmethyl) homo-octopine do not induce (47,91).

All cure and transfer experiments have shown that the following markers are borne on: 1) octopine Ti-plasmids -- oncogenicity (vir), induction of rough tumors synthesizing octopine (ocs), ability to catabolize octopine (occ), exclusion of phage AP1 (ape), self-transmissibility (tra); and 2) nopaline Ti-plasmids -- oncogenicity (vir), induction of smooth tumors synthesizing nopaline (nos), ability to catabolize nopaline (noc), exclusion of phage AP1 (ape), sensitivity to agrocin 84 (agr), self-transmissibility (tra). Plasmids lacking one or more of these markers have also been found in nature.

The ability to attach to plant cells can be determined not only by chromosomal genes but also by plasmid genes. Agrobacteria cured of their Ti-plasmid still attach to plant cells and may inhibit tumor induction by competition for attachment sites with virulent bacteria. For these strains, attachment is determined by chromosomal genes. A few nonpathogenic Agrobacteria have been isolated in nature that do not compete with pathogenic Agrobacteria for attachment sites and therefore probably lack the ability to attach. These strains, however, can be converted into pathogens by the introduction of an octopine or nopaline Ti-plasmid. The lipopolysaccharide of the converted strains does inhibit tumor induction by a pathogenic strain while the LPS of the original recipient does not. This indicates that genes located on the Ti-plasmid function in the LPS alterations that make attachment possible (92).

8. OPINE METABOLISM

The unusual amino acids found in crown gall tumors are generally referred to as opines. They are not essential for tumor induction since bacterial strains that induce tumors have been isolated in which these opines are absent. Agrobacteria that induce tumors containing octopine, octopinic acid, lysopine and histopine (Figure 3) are able to catabolize these same compounds, while bacteria that induce tumors containing nopaline and ornaline can degrade them (16,17,22,24; Table 1). Octopine strains do not consume nopaline or ornaline, while nopaline strains do not catabolize octopine, octopinic acid, lysopine or histopine. Morel (93) explained these results by assuming that during tumor induction bacterial DNA is introduced into the plant cell, is expressed, and codes for an enzyme that can catalyze both synthesis and degradation of the unusual amino acids. It is noteworthy that the octopine dehydrogenase of _Pecten maximum_ L. catalyses both synthesis and degradation.

The enzymes that catalyze the synthesis of octopine and nopaline in tumors have recently been purified to a considerable extent (94,95). They can also catalyze the degradation of octopine and nopaline. The octopine enzyme has been called lysopine dehydrogenase (LpDH) since lysine had the highest affinity for the enzyme in the presence of pyruvate as cosubstrate. _In vitro_ not only are lysine, ornithine, arginine and histidine substrates for LpDH, but so are methionine and glutamine, suggesting the possible existence of methiopine and glutaminopine in octopine tumors. Furthermore, in the presence of arginine as cosubstrate, pyruvate can be replaced as a substrate by α-ketobutyric acid, α-ketovaleric, α-ketocaproic acid and gly-oxylic acid but not by α-ketoglutaric acid (which would lead to nopaline synthesis). This suggests the possibility of many more as yet unknown unusual amino acids occurring in octopine tumors (95). For the nopaline enzyme (NpDH), only arginine and ornithine are substrates in the presence of α-ketoglutaric acid but not, for example,

Table 4

Similarities and Differences Between LpDH,
NpDH and Their Bacterial Counterparts

	Lysopine dehydrogenase[a]	Octopine oxidase[b]	Nopaline dehydrogenase[a]	Nopaline oxidase[b]
Octopine degradation	+	+	not tested	+
Nopaline degradation	not tested	–	+(weak)	+
Octopine synthesis	+	not tested	–	not tested
Nopaline synthesis	–	not tested	+	not tested
Location	cytoplasm	membrane-bound	cytoplasm	membrane-bound(?)
Coenzyme	NADPH		NADPH	

[a]From plant tumor.

[b]From A. tumefaciens.

lysine, histidine or methionine. Other α-ketoacids that may replace
α-ketoglutaric acid have not been found (96).
    Contrary to the hypothesis of Morel (93) differences have been
observed between the tumor-specific enzyme and the bacterial enzyme
(Table 4). The bacterial enzymes have been studied less thoroughly.
They are difficult to handle since, in contrast to the tumor-
specific enzyme which is soluble, they are membrane-bound, and possibly linked to the cytochromes (97). By toluenizing bacteria,
octopine oxidase activity has been measured and the enzyme has been
found to accept octopine, lysopine and octopinic acid as substrate
(87). Histopine, nopaline and ornaline have not been tested. Genetic evidence, however, shows that octopine, octopinic acid and lysopine are indeed substrates for octopine oxidase but nopaline is not
(47). The nopaline oxidase has been found to accept not only nopaline as substrate, but octopine, octopinic acid and lysopine, too
(15,98). Nopaline strains having the nopaline enzymes constitutively
also degrade octopine, octopinic acid and lysopine but octopine
strains with constitutive octopine enzymes do not catabolize nopaline
even when a nopaline permease is present in the cell (see below)
(15,47,98).
    Specific permeases, controlled by the same regulator as the
oxidases, are also involved in the catabolism of octopine and nopaline. The substrate specificity for the permeases is identical to
that of the oxidases (87,100). Octopine permease accepts octopine,

octopinic acid and lysopine as substrates but not arginine and nopaline, for example. Both nopaline and octopine serve as substrates for nopaline permeases (100).

Nopaline strains do not generally consume octopine, octopinic acid and lysopine. They do, however, when they are pregrown in media containing nopaline, indicating that the nopaline enzymes are inducible by nopaline (98). The octopine enzymes have been found to be inducible by octopine, octopinic acid and lysopine (100).

To test Morel's hypothesis (93) further, mutants of octopine strains that could no longer utilize octopine, octopinic acid and lysopine (99,101), and mutants of nopaline strains that could not catabolize nopaline (102) were isolated. Strains having a functional permease and apparently lacking the oxidase were shown to be virulent. They induced tumors that synthesized unusual amino acids. These data indicate that the bacterial enzyme and the tumor-specific enzyme are different.

In section 7 we reported that the transfer genes are inducible. Transfer of octopine Ti-plasmids can be induced by octopine, octopinic acid or lysopine and transfer of nopaline Ti-plasmids by nopaline. In fact, the transfer genes are inducible by the same compounds that also induce the catabolic enzymes (Table 5). Mutants have been isolated to check whether or not these genes are controlled by the same regulator, namely, 1) mutants constitutive for the catabolic enzymes (29,87,91), and 2) mutants derepressed for transfer (47,91).

1) Mutants constitutive for the octopine enzymes have been isolated as mutants able to catabolize an analogue of octopine that cannot induce the octopine genes, namely nor-octopine (Figure 4). These mutants are either constitutive or inducible by nor-octopine. Constitutive mutants are sensitive to (desmethyl) homo-octopine (Figure 4) another analogue that does not induce the octopine enzymes. (Desmethyl) homo-octopine is split into pyruvic acid (hence glyoxylic acid) and homo-arginine, which is toxic for Agrobacterium. Mutants that can no longer degrade octopine can therefore be isolated from constitutive mutants as bacteria able to grow in the presence of (desmethyl) homo-octopine.

2) Mutants derepressed for transfer have been isolated as strains having Ti-plasmids that are transferable in the absence, as well as the presence, of octopine.

The properties of mutants obtained have shown that the catabolic genes and the transfer genes are indeed controlled by the same regulator. Three types of mutants have been observed:

1) Bacteria constitutive for catabolism but inducible for transfer,

2) bacteria inducible for catabolism but derepressed for transfer, and

3) bacteria constitutive for catabolism and derepressed for transfer.

Types 1 and 2 probably are operator mutants, while type 3 probably have mutations in the regulator gene. Recently, mutants inducible

Table 5

Substrate Specificities of Octopine and Nopaline Catabolic
Enzymes, Regulator Genes and Transfer Genes

	Octopine permease	Octopine oxidase	Octopine plasmid transfer genes	Nopaline permease	Nopaline oxidase
INDUCTION BY:					
Octopine	+	+	+	−	NT[a]
Lysopine	+	+	+	−	NT[a]
Octopinic acid	+	+	+	−	NT[a]
Nopaline	−	NT[a]	−	+	+
ACCEPTANCE AS SUBSTRATE:					
Octopine	+	+		+	+
Lysopine	+	+		+	+
Octopinic acid	+	+		+	+
Nopaline	−	−		+	+

[a]NT, not tested, probably negative if oxidase and permease are in
one operon.

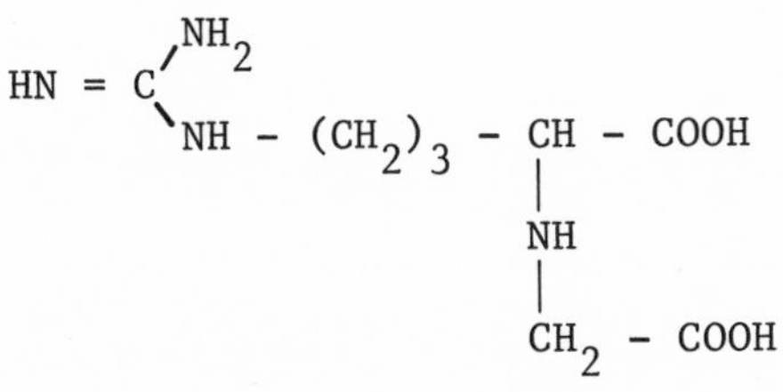

$$HN = C \begin{smallmatrix} \nearrow NH_2 \\ \searrow NH \end{smallmatrix} - (CH_2)_3 - CH - COOH$$
$$| \\ NH \\ | \\ CH_2 - COOH$$

Nor-octopine

$$HN = C \begin{smallmatrix} \nearrow NH_2 \\ \searrow NH \end{smallmatrix} - (CH_2)_4 - CH - COOH$$
$$| \\ NH \\ | \\ CH_3 - CH - COOH$$

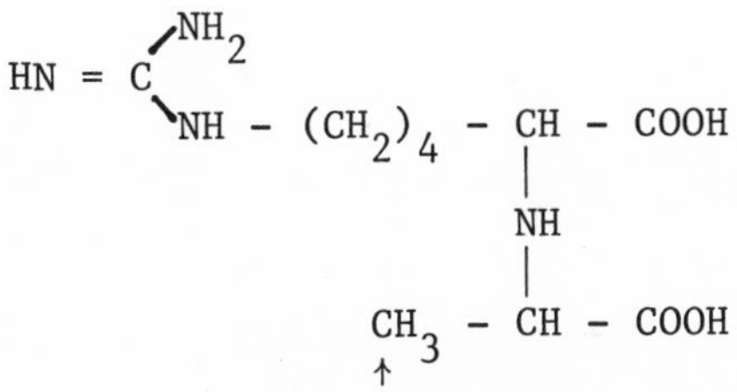

Homo-octopine

Figure 4. Octopine analogues used for the isolation of mutant plasmids. (Desmethyl) homo-octopine lacks the methyl group indicated by
the arrow.

with nor-octopine for the octopine genes have been observed to be
also inducible by nor-octopine for transfer. These data are in
perfect agreement with the model above.

   It is very remarkable that Ti-plasmid transfer becomes derepressed in the presence of compounds that can be consumed by
the host strain with plasmid-coded enzymes, leading to the spread

of the plasmid in this niche. The same may be true for more degradative plasmids. For one plasmid in particular, the <u>Pseudomonas</u>
plasmid, regulation of transfer and catabolism may be linked. Mutants partially derepressed for transfer have been shown to contain
higher levels of catabolic enzymes (103).

9.   MAP OF AN OCTOPINE Ti-PLASMID

     Restriction enzyme maps have been constructed for the octopine
Ti-plasmid of strain B6-806 and the nopaline Ti-plasmid of strain
C58 (104,105). By comparing the restriction enzyme patterns of
deletion or insertion mutants of Ti-plasmids with those of the
parental Ti-plasmid, genes can be localized on the Ti-plasmid map.
Deletion mutants have been isolated as nonreverting homo-octopine
resistant mutants from a B6 octopine Ti-plasmid that acquired Tn1
from RP4 and has become derepressed for transfer and catabolic
enzymes (106). The majority of the deletions were found to begin
in the vicinity of Tn1; deletions frequently start at the ends of
transposons (107). Deletion mutants have also been isolated in
the same manner from a similar plasmid lacking Tn1 (106). This
gave the genetic map shown in Figure 5 (106).

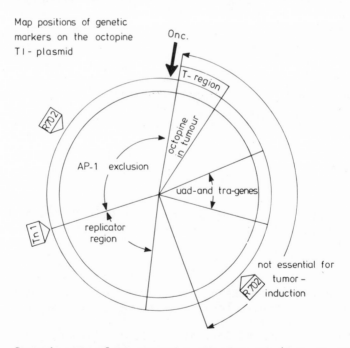

Figure 5. Genetic map of an octopine Ti-plasmid (from B. Koekman
et al. (1978), Plasmid, in press). Two insertion sites of the Inc P
plasmid R702 are indicated as is the position of a Tn1 insertion.

The genes involved in octopine catabolism are located close to the genes coding for octopine synthesis in tumors. Therefore, they may have originated from the same gene by gene duplication (see section 8, above).

Strains with plasmids lacking part or all of their T-DNA (the DNA region present in plant tumor tissue) are still (weakly) tumorigenic, but they induce tumors lacking LpDH activity (G. Ooms and B. Koekman, unpublished results). It may be that a small region close to this T-DNA contains the real oncogene (see the arrow in Figure 5), and that the bacterial DNA remaining in the plant tumor cells does not have fixed ends but is closely linked to the small oncogene.

Fine mapping is being performed by introducing transposons into the Ti-plasmid; these prevent expression of the gene into which they insert. One disadvantage is that they may have polar effects on distal genes. Insertions can be mapped by restriction enzyme analysis. Transposons have been introduced into Ti-plasmids in various ways.

1) Ti-plasmids with a transposon on the R-plasmid have been obtained by mobilization of Ti-plasmids with R-plasmids (see section 7, above) (86).

2) Transposons have been inserted by transferring transfer-negative R-plasmids into Ti-plasmid-harboring strains, either by transformation or by the use of R-plasmids containing phage Mu. The latter can be transferred by conjugation into Agrobacterium. Exconjugants have been obtained with low frequency that harbor deleted, and often tra⁻, R-plasmids lacking the Mu insertion. On very rare occasions, a complete plasmid containing Mu is transferred (8,108,109). Strains having a tra⁻ R-plasmid and a Ti-plasmid can be used as donors; exconjugants selected for an R-plasmid resistance marker either contain a Ti-plasmid plus the mobilized tra⁻ R-plasmid or a Ti-plasmid with a transposon from the R-plasmid.

3) Other R-plasmid::Mu cointegrates cannot establish themselves at all in Agrobacterium. If a transposon is present, it may be directly selected after conjugation. Frequently, transposons have been found to integrate into the chromosome at this time. Strains having a Ti-plasmid and a transposon in the chromosome have been used as donors; recipients selected for the resistance marker(s) of the transposon harbor Ti-plasmids containing it (110,111).

4) Transposon-containing cointegrate plasmids have been obtained by transferring a cointegrate plasmid that consists of RP4 and a Ti-plasmid into strains of E. coli that have transposons in their chromosomes. These have been transferred back to Agrobacterium in order to examine phenotypic alterations in Ti-plasmid markers (112). The Ti-plasmid genes are not expressed in E. coli (88,112).

With the use of transposon-containing Ti-plasmids, Ti-plasmid markers have been mapped. Some avirulent mutants have been isolated in which the transposon was distal to the T-DNA on the left side of the plasmid (Figure 5), suggesting that the genes involved in oncogenicity are on the left side of the plasmid.

Table 6

Surface Exclusion and Incompatibility Between
Inc P Plasmids RP4 and R702 in A. tumefaciens

Transferred plasmid	Plasmid in recipient	Transfer frequency	Original resistance markers (recipient)	Resistance markers (exconjugants)
RP4[a]	–	10^{-2}	–	CbKmTc
RP4	R702	2×10^{-4}[c]	SmSpKmTc	CbKmTc
R702[b]	–	4×10^{-3}	–	SmSpKmTc
R702	RP4	4×10^{-5}[c]	CbKmTc	SmSpKmTc

[a]RP4, CbKmTcIncP.

[b]R702, SmSpKmTcIncP.

[c]This very low transfer frequency is indicative of surface exclusion.
Abbreviations: Sm, streptomycin; Sp, spectinomycin; Km, kanamycin;
Tc, tetracycline; Cb, carbenicillin.

10.   INCOMPATIBILITY BETWEEN Agrobacterium PLASMIDS

Plasmid incompatibility is the inability of two different
plasmids to coexist stably, in the same host cell, in the absence
of continued selection pressure (113). Inc P plasmids that are
incompatible with E. coli and Pseudomonas aeruginosa are also
incompatible with Agrobacterium (Table 6), indicating that the genes
involved in incompatibility are expressed in Agrobacterium (86).

Agrobacteria isolated from nature have been found to carry
either a nopaline or an octopine Ti-plasmid but not both, suggesting
that these plasmids belong to the same incompatibility group. Incompatible plasmids usually share DNA homology (114); octopine and
nopaline Ti-plasmids only share a small region of strong DNA homology, which may contain the DNA essential for oncogenicity (58,115,
116). The restriction enzyme patterns of octopine and nopaline
Ti-plasmids share only one common band. Octopine Ti-plasmids of
different origin have almost identical restriction patterns, but
the nopaline Ti-plasmids are very different from each other (75).
Recently, large regions of weaker DNA homology have been found
between nopaline and octopine Ti-plasmids (115,116).

Incompatibility between different Ti-plasmids has been tested
by introduction of an octopine Ti-plasmid carrying Tnl into a
nopaline strain (117). Recipients that became resistant to carbenicillin and gained the ability to utilize octopine, lost the
ability to utilize nopaline, showing that octopine and nopaline
Ti-plasmids do indeed belong to the same incompatibility group

Table 7

Incompatibility Between Agrobacterium Plasmids

Transferred plasmid	Plasmid in recipient	Relevant plasmid markers	Relevant exconjugant markers
pAL208[a]	-	-	Cb
	pTiB6	occ$^+$[b]	Cb(occ$^-$)
	pTiC58	noc$^+$[c]	Cb(noc$^-$)
	pAtAG60	occ$^+$	Cb occ$^+$
	pAtKerr14	noc$^+$	Cb noc$^+$

[a] pAL208 is derived from pAL657 (pTiB6::Tn1). It differs by having a small deletion in the genes for octopine catabolism.

[b] occ, octopine catabolism.

[c] noc, nopaline catabolism.

(Table 7). By using a derivative of the above mentioned pTiB6::Tn1 plasmid containing a small deletion in the genes coding for octopine utilization, it was shown that octopine Ti-plasmids of different origin (B6, NCPPB4, 147) are incompatible with each other. Recipients which became carbenicillin-resistant, invariably lost their octopine utilization trait.

A transmissible plasmid from A. tumefaciens AG60 (8) capable of conferring octopine utilization on its host cell, but not AP1 exclusion or oncogenicity, has been found to be compatible with the Ti-plasmid (Table 7). This plasmid may be of a different origin and may have picked up the genes for octopine utilization from a Ti-plasmid by recombination or transposition. A plasmid from strain Kerr 14 that confers the ability to utilize nopaline and Kerr 84 sensitivity but not AP1 exclusion or oncogenicity, has also been shown to be compatible with Ti-plasmids (Table 7) (117).

Crosses with octopine Ti-plasmid-containing strains as donors and nopaline Ti-plasmid-harboring strains as recipients result, with low frequency, in exconjugants able to break down nopaline in spite of the introduction of the octopine Ti-plasmid (117). These strains carry recombinant plasmids consisting of a complete octopine and a complete nopaline plasmid. The properties of such a strain are listed in Table 8. As can be seen in this table, tumors induced by these strains are smooth and contain nopaline but not octopine, indicating a predominance of the nopaline Ti portion of the cointegrate. However, despite the fact that octopine is not detectable, the enzyme LpDH that catalyzes its synthesis (117) can be demonstrated. This indicates that the octopine Ti portion

Table 8

Cointegrate-Plasmid-Harboring Strain Compared With Donor and Recipient Strains

| | A. tumefaciens | | |
| | Str resistant[c] | Rif resistant[d] | |
	LBA 696	LBA 670	LBA 298
Phage sensitivity:			
S2, S5, S6	+	+	+
S1, S3	−	−	−
Ti-plasmid type	octopine	cointegrate	nopaline
Plasmid size[a]	120	>200	120
Octopine utilization	+	+	−
Nopaline utilization	−	+	+
Phage AP1 exclusion	+	+	+
Agrocin 84 sensitivity	−	+	+
Tumor induction	+	+	+
Tumor morphology (Kal- anchoë diagremontiana)	rough	smooth	smooth
LpDH activity[b]	+	+	−
NpDH activity[b]	−	+	+

[a]In megadaltons.

[b]In tumors.

[c]Str, streptomycin.

[d]Rif, rifampicin.

is active in tumorigenesis, also. The cointegrate plasmids are
stable -- cotransfer of all markers is 100%, mutants resistant to
agrocin 84 or homo-octopine still harbor plasmids, although they
have small deletions or point mutations, and the plasmids are maintained, even when the bacteria are grown at 37°C. This indicates
that the plasmids do not tend to lose either the octopine or the
nopaline Ti-plasmid and that reversion of the recombination process,
which leads to the formation of these cointegrates, does not occur
with detectable frequency.

11.   HOST RANGE OF Ti-PLASMIDS

Ti-plasmids cannot be maintained by E. coli (88). Even RP4::Ti
cointegrate plasmids appear to be unstable in this host and yield

Table 9

Properties of an R. trifolii Strain and Exconjugants
Harboring Various Ti-Plasmids

Ti-plasmids were transferred from various A. tumefaciens strains
into R. trifolii LPR 5002, a streptomycin- and rifampicin-resistant
derivative of strain RT5.

	R. trifolii (Str-, Rif-resistant)			
	LPR 5002	LPR 511	LPR 518	LPR 519
Phage sensitivity				
LPB1	+	+	+	+
LPB51	+	+	+	+
Ti-plasmid	−	+	+	+
Ti-plasmid type		oct	oct::nop	nop[d]
Octopine breakdown	−	+	+	+
Nopaline breakdown	−	−	+	+
Tumor induction[a]	−	+	+	+
Tumor morphology[b]		rough	smooth	smooth
LpDH activity[c]		+	+	−
NpDH activity[c]		−	+	+
Nodulation of:				
Trifolium pratense	+	+	+	+
Trifolium parviflorum	+	+	+	+
Nitrogen-fixation				
in nodules	+	+	+	+

[a] In Kalanchoë diagremontiana and Helianthus annuus.

[b] In Kalanchoë diagremontiana.

[c] In tumors.

[d] Constitutive.

strains that have lost the Ti portion of the cointegrate. Neither
the octopine utilization trait nor virulence are expressed in
E. coli (88,112).
    However, it is possible to introduce Ti-plasmids into another
member of the family Rhizobiaceae, Rhizobium trifolii (15,62).
Moreover, the Ti-plasmid is fully expressed in R. trifolii (Table 9).
Depending on the plasmid received, R. trifolii induces either
smooth tumors containing nopaline or rough tumors containing octopine. Ti-plasmids are stably maintained in Rhizobium. Even cointegrates consisting of an octopine Ti-plasmid and a nopaline Ti-
plasmid can be transferred to R. trifolii. Like all other Ti-plasmids,

these cointegrates are stably maintained in this host. Ti-plasmids
can be transferred back from Rhizobium to Agrobacterium. Transfer
of wild-type octopine Ti-plasmids in Rhizobium takes place only
in the presence of octopine. Nopaline plasmids have not been
tested for this. Agrobacterium strains receiving a Ti-plasmid
from Rhizobium did not at the same time acquire from Rhizobium the
ability to induce nodules.
     R. trifolii strains harboring a Ti-plasmid are still able to
nodulate their proper hosts (Trifolium species) effectively.
Neither octopine nor nopaline have ever been detected in nodules,
while tumors induced by these same strains have never shown
nitrogen-fixation. This indicates that the processes of nodulation
and tumorigenesis can be induced by one and the same organism,
though they are completely separate processes.
     The large plasmids that are naturally present in R. trifolii
remain in the presence of a Ti-plasmid, showing the compatibility
between these plasmids (62). DNA homology studies have not shown
any homology between octopine Ti-plasmids and these Rhizobium
plasmids (118).

     12. PROSPECTS FOR THE GENETIC ENGINEERING OF PLANTS

     Since at least part of a bacterial plasmid is transferred into
plant cells by A. tumefaciens in the process of tumor induction
(76-78), and since this T-DNA (bacterial DNA found in transformed
plant tissue) does not have fixed ends and therefore cannot be a
transposon (106), then, in principle, genetic information may be
intentionally introduced into plant genomes. It has already been
shown that foreign DNA can enter plant cells in this way. Plant
cells transformed with a strain carrying a Ti-plasmid with an
insertion of Tn7 in its T-region, have been found to contain this
Tn7 DNA (119). Therefore, A. tumefaciens cells provided with the
desired genetic information at the right position within their
plasmid may possibly be used in the future to achieve a reproducible
and efficient transfer of genes into plant cells. An in vitro
system for transformation of plant protoplasts with A. tumefaciens
cells has been developed whereby numerous different transformed
tissues have been obtained containing LpDH activity (the tumor-
specific marker) (120).
     Foreign DNA may be cloned in Agrobacterium as soon as cloning
vehicles are available for this organism. A derivative of a Ti-
plasmid that has lost most of its DNA by deletion and has a molecular weight of about 25 million, or an even smaller plasmid derived
from this plasmid in vitro, may be used as such a cloning vehicle
(106).
     Among the genes whose introduction into plant cells would be
useful are genes from other plant species, to obtain hybrids, and
nitrogen-fixation genes. The nif genes of Rhizobium are probably
located on a large plasmid (121), so their introduction into

Agrobacterium may be possible by conjugation. However, this would probably not yield bacteria with a wider host range for symbiotic nitrogen-fixation since in R. trifolii strains carrying a Ti-plasmid, tumorigenesis and nodulation are fully separated processes. These strains do not induce nitrogen-fixing tumors, not do they have a wider host range for symbiotic nitrogen-fixation (8,15). The host range of Rhizobia may be extended, however, if the genes for certain legume lectins -- the enzymes that are probably recognized by their Rhizobium symbiont (122-124) -- are introduced into new plant species. This may be realized after the DNA coding for the lectin has been cloned in a Ti-plasmid. However, such plants may fix nitrogen symbiotically only after genes coding for leghemoglobin synthesis have been transferred into them in the same way.

Acknowledgment: We gratefully acknowledge the help of our colleagues, Drs. Klapwijk, Koekman, Ooms, Otten, Wullems and Würzer-Figurelli, in preparing this manuscript.

REFERENCES

1   Smith, E.F. and Townsend, C.O. (1907) Science 25, 671.
2   Panagopoulos, C.G. and Psallidas, P.G. (1973) J. Appl. Bacteriol. 36, 233.
3   Süle, S. (1978) J. Appl. Bacteriol. 44, 207.
4   Kerr, A. (1972). J. Appl. Bacteriol. 35, 493.
5   Tamm, B. (1954) Arch. Mikrobiol. 20, 273.
6   de Cleene, M. and de Ley, J. (1977) Bot. Rev. 42, 389.
7   Kado, C.I. (1978) EMBO Workshop on Plant Tumor Research, Noordwijkerhout, The Netherlands.
8   Hooykaas, P.J.J. (unpublished results).
9   Lippincott, J.A. and Lippincott, B.B. (1975) Ann. Rev. Microbiol. 29, 377.
10 Bopp, M. (1966) Biol. Rundsch. 4, 25.
11 Lippincott, B.B. and Lippincott, J.A. (1969).J. Bacteriol. 97, 620.
12 Schilperoort, R.A. (1969) Ph.D. Dissertation, Leiden, The Netherlands.
13 Whatley, M.H., Bodwin, J.A., Lippincott, B.B. and Lippincott, J.A. (1976) Infect. Immun. 13, 1080.
14 Schilperoort, R.A., Kester, H.C.M., Klapwijk, P.M., Rörsch, A. and Schell, J. (1975) Communication Semaine de'Etudes Agr. Hyg. Plantes, Gembloux, France.
15 Hooykaas, P.J.J., Klapwijk, P.M., Nuti, M.P., Schilperoort, R.A. and Rörsch, A. (1977) J. Gen. Microbiol. 98, 477.
16 Petit, A., Delhaye, S. Tempé, J. and Morel, G. (1970) Physiol. Vég. 8, 205.
17 Bomhoff, G., Klapwijk, P.M., Kester, H.C.M., Schilperoort, R.A., Hernalsteens, J.P. and Schell, J. (1976). Mol. Gen. Genet.145, 177.

18 Ménagé, A. and Morel, G. (1964) C.R. Acad. Sci. Paris sér. D 259, 4795.

19 Ménagé, A. and Morel, G. (1965) C.R. Acad. Sci. Paris sér. D 261, 2001.

20 Lioret, C. (1956) Physiol. Vég. 2, 76.

21 Beimann, K., Lioret, C., Asselineau, K., Lederer, E. and Polonski, T. (1960) Bull. Soc. Chim. Biol. 42, 979.

22 Kemp, J.D. (1977) Biochem. Biophys. Res. Commun. 74, 862.

23 Goldmann, A., Thomas, D.W. and Morel, G. (1969) C.R. Acad. Sci. Paris sér. D 268, 852.

24 Firmin, J.L. and Fenwick, R.G. (1977) Phytochemistry 16, 761.

25 Kemp, J.D., Hack, E., Sutton, D.W. and El-Wakil, M. (1978) EMBO Workshop on Plant Tumor Research, Noordwijkerhout, The Netherlands.

26 Bergey's Manual of Determinative Bacteriology, 7th Edition (1957) (Breed, R.S., Murray, E.G.D. and Smith, N.R., eds.), The Williams and Wilkins Co., Baltimore, MD.

27 Graham, P.H. (1964) J. Gen. Microbiol. 35, 511.

28 Kersters, K., de Ley, J., Sneath, P.H.A. and Sackin, M. (1973) J. Gen. Microbiol. 78, 227.

29 Moffett, M.L. and Colwell, R.R. (1968) J. Gen. Microbiol. 51, 245.

30 White, L.O. (1972) J. Gen. Microbiol. 72, 565.

31 Kerr, A. and Panagopoulos, C.G. (1977) Phytopathol. Z. 90, 172.

32 MacGregor, A.N. and Alexander, M. (1971) J. Bacteriol. 105, 728.

33 Skinner, F.A. (1977) J. Appl. Bacteriol. 43, 91.

34 Keane, P.J., Kerr, A. and New, P.B. (1970) Aust. J. Biol. Sci. 23, 585.

35 Gibbins, A.M. and Gregory, K.F. (1972) J. Bacteriol. 111, 129.

36 Lippincott, J.A., Beiderbeck, R. and Lippincott, B.B. (1973) J. Bacteriol. 116, 378.

37 Watson, B., Currier, T.C., Gordon, M.P., Chilton, M.-D. and Nester, E.W. (1975) J. Bacteriol. 123, 255.

38 Langley, R.A. and Kado, C.I. (1972) Mutat. Res. 14, 277.

39 Klapwijk, P.M., de Jonge, A.J.R., Schilperoort, R.A. and Rörsch, A. (1975) J. Gen. Microbiol. 91, 177.

40 Schilde-Rentschler, L., Gordon, M.P., Saiki, R. and Melchers, G. (1977) Mol. Gen. Genet. 155, 235.

41 Beringer, J.E. (1974) J. Gen. Microbiol. 84, 188.

42 Klapwijk, P.M., Hooykaas, P.J.J., Kester, H.C.M., Schilperoort, R.A. and Rörsch, A. (1976) J. Gen. Microbiol. 96, 155.

43 Holsters, M., de Waele, D., de Picker, A., Messens, E., van Montagu, M. and Schell, J. (1978) Mol. Gen. Genet. 163, 181.

44 O'Gara, F. and Dunican, L.K. (1973) J. Bacteriol. 116, 1177.

45 Kowalski, M. (1967) Acta Microbiol. Pol. 16, 7.

46 Svab, Z., Kondorosi, A. and Orosz, L. (1978) J. Gen. Microbiol. 106, 321.

47 Hooykaas, P.J.J., Roobol, C. and Schilperoort, R.A. (1978) J. Gen. Microbiol. (in press).

48   Datta, N. and Hedges, R.W. (1972) J. Gen. Microbiol. 70, 453.
49   Hernalsteens, J.P., Villaroel-Mandiola, R., van Montagu, M. and
      Schell, J. (1977) in DNA Insertion Elements, Plasmids and
      Episomes (Bukhari, A.I., Shapiro, J.A. and Adhya, S.L., eds.),
      pp. 179-183, Cold Spring Harbor Laboratory, Cold Spring Harbor,
      NY.
50   Cole, M.A. and Elkan, G.H. (1973) Antimicrobiol. Agents
      Chemother. 4, 248.
51   Kondorosi, A., Kiss, G.B., Forrai, T., Vincze, E. and Banfalvi,
      Z. (1977) Nature 268, 525.
52   Meade, H.M. and Signer, E.R. (1977) Proc. Nat. Acad. Sci.
      U.S.A. 74, 2076.
53   Beringer, J.E. and Hopwood, D.A. (1976) Nature 264, 291.
54   Beringer, J.E., Hoggan,S.A. and Johnston, A.W.B. (1978)
      J. Gen. Microbiol. 104, 201.
55   Johnston, A.W.B. and Beringer, J.E. (1977) Nature 267, 611.
56   Johnston, A.W.B., Setchell, S.M. and Beringer, J.E. (1978)
      J. Gen. Microbiol. 104, 209.
57   Zaenen, I., van Larebeke, N., Teuchy, H., van Montagu, M. and
      Schell, J. (1974) J. Mol. Biol. 86, 109.
58   Currier, T.C. and Nester, E.W. (1976) J. Bacteriol. 126, 157.
59   Merlo, D.J. and Nester, E.W. (1977) J. Bacteriol. 129, 76.
60   Sheikholeslam, S., Okubara, P.A., Lin, B.-C., Dutra, J.C. and
      Kado, C.I. (1978) Microbiology (in press).
61   Nuti, M.P., Ledeboer, A.M., Lepidi, A.A. and Schilperoort, R.A.
      (1977) J. Gen. Microbiol. 100, 241.
62   Prakash, R.K., Hooykaas, P.J.J., Ledeboer, A.M., Kijne, J.,
      Schilperoort, R.A., Nuti, M.P., Lepidi, A.A., Casse, F.,
      Boucher, C., Julliot, J.S. and Dénarié, J. (1978) Proc. 3rd
      Int. Symp. Nitrogen Fixation, Madison, WI.
63   Casse, F., Boucher, C., Julliot, J.S. and Dénarié, J. (1978)
      J. Gen. Microbiol. (in press).
64   Currier, T.C. and Nester, E.W. (1976) Anal. Biochem. 76, 431.
65   Ledeboer, A.M., Krol, A.J.M., Dons, J.J.M., Spier, F.,
      Schilperoort, R.A., Zaenen, I., van Larebeke, N. and Schell, J.
      (1976) Nucl. Acids Res. 3, 449.
66   Clewell, D.B. and Helinski, D.R. (1969) Proc. Nat. Acad. Sci.
      U.S.A. 62, 1159
67   Dunican, L.K., O'Gara, F. and Tierney, A.B. (1976) in
      Symbiotic Nitrogen Fixation in Plants (Nutman, P.S., ed.),
      Cambridge Univ. Press, Cambridge, England.
68   van Larebeke, N., Engler, G., Holsters, M., van den Elsacker, S.,
      Zaenen, I., Schilperoort, R.A. and Schell, J. (1974) Nature
      252, 169.
69   Lin, B.-C. and Kado, C.I. (1977) Can. J. Microbiol. 23, 1554.
70   Hamilton, R.H. and Fall, M.Z. (1971) Experientia 27, 229.
71   Kerr, A. (1971) Physiol. Plant Pathol. 1, 241.
72   Kerr, A. and Roberts, W.P. (1976) Physiol. Plant Pathol. 9, 205.
73   Albinger, G. and Beiderbeck, R. (1977) Phytopathol. Z. 90, 306.

74   van Larebeke, N., Genetello, C., Schell, J., Schilperoort, R.A.,
     Hermans, A.K., Hernalsteens, J.P. and van Montagu, M. (1975)
     Nature 255, 742.
75   Sciaky, D., Montoya, A.L. and Chilton, M.-D. (1978) Plasmid
     1, 238.
76   Chilton, M.-D., Drummond, M.H., Merlo, D.J., Sciaky, D.,
     Montoya, A.L., Gordon, M.P. and Nester, E.W. (1977) Cell 11, 263.
77   Drummond, M.H., Gordon, M.P., Nester, E.W. and Chilton, M.-D.
     (1977) Nature 269, 535.
78   Ledeboer, A.M. (1978) Ph.D. Dissertation, Leiden, The
     Netherlands.
79   Roberts, W.P. and Kerr, A. (1974) Physiol. Plant Pathol. 4, 81.
80   Engler, G., Holsters, M., van Montagu, M., Schell, J.,
     Hernalsteens, J.P. and Schilperoort, R.A. (1975) Mol. Gen.
     Genet. 138, 345.
81   Kerr, A. and Htay, K. (1974) Physiol. Plant Pathol. 4, 37.
82   van Larebeke, N., Genetello, C., Hernalsteens, J.P., de Picker,
     A., Zaenen, I., Messens, E., van Montagu, M. and Schell, J.
     (1977) Mol. Gen. Genet. 152, 119.
83   Hamilton, R.H. and Chopan, M.N. (1975) Biochem. Biophys. Res.
     Commun. 63, 349.
84   Bomhoff, G., Klapwijk, P.M., Kester, H.C.M., Schilperoort, R.A.,
     Hernalsteens, J.P. and Schell, J. (1976) Mol. Gen. Genet. 145,
     177.
85   Chilton, M.-D., Farrand, S.K., Levin, R. and Nester, E.W.
     (1976) Genetics 83, 609.
86   Hooykaas, P.J.J., den Dulk-Ras, A., Foekens, J. and Schilperoort,
     R.A. (unpublished results).
87   Klapwijk, P.M., Scheulderman, T. and Schilperoort, R.A. (1978)
     J. Bacteriol. (in press).
88   Holsters, M., Silva, B., van Vliet, F., Hernalsteens, J.P.,
     Genetello, C., van Montagu, M. and Schell, J. (1978) Nature
     (in press).
89   Kerr, A., Manigault, P. and Tempé, J. (1977) Nature 265, 560.
90   Genetello, C., van Larebeke, N., Holsters, M., de Picker, A.
     van Montagu, M. and Schell, J. (1977) Nature 265, 561.
91   Petit, A., Tempé, J., Kerr, A., Holsters, M., van Montagu, M.
     and Schell, J. (1978) Nature 271, 570.
92   Whatley, M.H., Margot, J.B., Schell, J., Lippincott, B.B. and
     Lippincott, J.A. (1978) J. Gen. Microbiol. 107, 395.
93   Morel, G. (1971) Coll. Int. CNRS (INRA, ed.) 193, 463.
94   Otten, L.A.B.M., Vreugdenhil, D. and Schilperoort, R.A. (1977)
     Biochim. Biophys. Acta 485, 268.
95   Otten, L., Kester, H. and Schilperoort, R.A. (1978) EMBO
     Workshop on Plant Tumor Research, Noordwijkerhout, The
     Netherlands.
96   Otten, L.A.B.M. (unpublished results).
97   Bomhoff, G.H. (1974) Ph.D. Dissertation, Leiden, The
     Netherlands.
98   Petit, A. and Tempé, J. (1975) C.R. Acad. Sci. Paris 281, 467.

99   Petit, A. and Tempé, J. (1976) C.R. Acad. Sci. Paris 282, 69.
100 Klapwijk, P.M., Oudshoorn, M. and Schilperoort, R.A. (1977)
     J. Gen. Microbiol. 102, 1.
101 Klapwijk, P.M., Hooykaas, P.J.J., Kester, H.C.M., Schilperoort,
     R.A. and Rörsch, A. (1976) J. Gen. Microbiol. 96, 155.
102 Montoya, A.L., Chilton, M.-D., Gordon, M.P., Sciaky, D. and
     Nester, E.W. (1977) J. Bacteriol. 129, 101.
103 Nakazawa, T. and Yokota, T. (1977) J. Bacteriol. 129, 39.
104 Chilton, M.-D., Montoya, A.L., Merlo, D.J., Drummond, M.H.,
     Nutter, R., Gordon, M.P. and Nester, E.W. (1978) Plasmid 1,
     254.
105 Wilde, M.D., de Vos, G., van Montagu, M. and Schell, J. (1978)
     EMBO Workshop on Plant Tumor Research, Noordwijkerhout, The
     Netherlands.
106 Koekman, B.P., Ooms, G., Klapwijk, P.M. and Schilperoort, R.A.
     (1978) Plasmid (in press).
107 Kopecko, D.J., Brevet, J. and Cohen, S.N. (1976) J. Mol. Biol.
     108, 333.
108 Klapwijk, P.M., van Breukelen, J. and Schilperoort, R.A.
     (in preparation).
109 Boucher, C. Bergeron, B., Bertalmio, M.B.D. and Dénarié, J.
     (1977) J. Gen. Microbiol. 98, 253.
110. Schell, J. and van Montagu, M. (personal communication).
111 Beringer, J.E. (personal communication).
112 Hernalsteens, J.P., de Greve, H., van Montagu, M. and Schell,
     J. (1978) Plasmid 1, 218.
113 Novick, R.P., Clowes, R.C., Cohen, S.N., Curtiss, R. III,
     Datta, N. and Falkow, S. (1976) Bacteriol. Rev. 40, 168.
114 Grindley, N.D.F., Humphreys, G.D. and Anderson, E.S. (1973)
     J. Bacteriol. 115, 387.
115 de Picker, A., van Montagu, M. and Schell, J. (1978) Nature
     275, 150.
116 Chilton, M.-D., Drummond, M.H., Merlo, D.J. and Sciaky, D.
     (1978) Nature 275, 147.
117 Hooykaas, P.J.J., den Dulk-Ras, A. and Schilperoort, R.A.
     (in preparation).
118 Prakash, R.K. (personal communication).
119 Hernalsteens, J.P., van Montagu, M. and Schell, J. (1978)
     EMBO Workshop on Plant Tumor Research, Noorwijkerhout, The
     Netherlands.
120 Márton, L., Wullems, G., Molendijk, L. and Schilperoort, R.A.
     (1978) Nature (in press).
121 Nuti, M.P., Cannon, F.C., Prakash, R.K. and Schilperoort, R.A.
     (1978) Nature (in press).
122 Dazzo, R.B. and Hubbell, B.H. (1975) Appl. Microbiol. 30, 1017.
123 Wolpert, J.S. and Albersheim, P. (1976) Biochem. Biophys. Res.
     Commun. 70, 729.
124 Planqué, K. and Kijne, J. (1977) FEBS Lett. 73, 59.

THE CHLOROPLAST, ITS GENOME AND POSSIBILITIES FOR GENETICALLY

MANIPULATING PLANTS

L. Bogorad

The Biological Laboratories
Harvard University
16 Divinity Avenue
Cambridge, Massachusetts   02138

INTRODUCTION

We can imagine altering the genetic potential of an organism
by deleting or introducing structural genes. We can visualize
altering patterns of gene expression by manipulating regulatory DNA
sequences. Exercise of either of these two genetic engineering
approaches will reveal important biological principles and may well
lead to important applications.

Today, we induce mutations randomly and select organisms with
those alterations we wish for our studies. Or, by adjusting an
organism's environment, we influence gene expression to learn about
the organization of a genome and mechanisms of controlling gene
expression. Tomorrow, genetic engineering techniques should enable
us to see the effects of controlled perturbations and thus, through
in vitro initiated genetic changes, lead us to recognize more
directly how cells work. Genetic engineering of eukaryotic plants
holds the promise of special benefits--the possibility of modifying
cells or whole plants for crop improvement or industrial processes.

Each cell of a higher green plant has not only nuclear and
mitochondrial genomes, like other eukaryotic cells, but its plastids
have a third unique genome. Plastids offer special challenges and
special opportunities for genetically engineered structural and
regulatory changes. To understand what may be possible, it is useful to review our small store of knowledge of plastid genomes and to
recall the characteristics of some differentiated plastid types.

Soon after the beginning of the century, when it was already
known that some plastid characters were transmitted according to
the Mendelian rules of inheritance, it was realized that some other
characters that affect chloroplast development were not transmitted

181

according to these rules (1,2). Those features inherited according to the rules of Mendel were correctly taken to be controlled by nuclear genes. Those transmitted according to some other system were judged to be the products of "plastome" genes. Thus, long before chloroplasts and other plastids were shown to contain DNA in the early and mid-1960s (3), the notion of their containing genetic information was well established among plant geneticists.

Between the early 1960s and the early to mid-1970s, it became well established that plastids contain unique equipment for the storage and expression of genetic information (3). Their DNA is grossly different from that of the nucleus. Their ribosomes differ in size, protein and RNA composition from those of the cytoplasm. Chloroplast DNA-dependent RNA polymerase is different from that of the nucleus, at least in maize. From transmission genetics, we know of genetic markers for antibiotic resistance in the plastid genome of the single-celled Chlamydomonas reinhardi. Finally, a few structural genes have been mapped physically on the maize plastid chromosome. An outline of these data comprises the first part of the ensuing discussion. Research ahead will undoubtedly define further the organization of the chloroplast genome and provide an understanding of the rules of gene exchange and other characteristics of the uniquely evolved plastid genome, a genome that is likely to differ in some fundamental general properties from the genomes of prokaryotes, nuclei and even mitochondria.

Plastids can differentiate into many specialized forms: green photosynthetic chloroplasts; red or orange carotenoid-containing chromoplasts, e.g. in ripe tomatoes; starch-storing amyloplasts; and oil-containing elaioplasts (4). Chloroplast development is arrested at the etioplast stage in dark-grown seedlings. Chloroplasts in mesophyll and bundle sheath cells in leaves of tropical grasses and other C4 plants develop into two functionally different types. Thus these organelles provide many situations for studying the control of gene expression. Genetic engineering may hasten our understanding of these controls and designed alterations in patterns of differentiation may lead to the production of agriculturally and industrially desirable cells or whole organisms.

Genetic engineering experiments require adequate amounts of purified, well characterized DNA sequences: structural genes, fragments of genes, regulatory genes, control elements adjacent to genes, etc. The potential recipient genome's organization, gene expression control systems and replication mechanisms all need to be understood for intelligent stable genetic engineering to succeed but genetic engineering technology will itself facilitate understanding the recipient's genome. Another requirement is a cell at a stage suitable for taking up new genetic material and integrating it into the replicating system of the genome. Of all plastid DNAs, most is known about that of Zea mays. Recognition sites for some restriction endonucleases have been mapped and a few genes have been located physically on the chromosome. Segments comprising

most of the maize plastid chromosome have been introduced into
plasmids cloned in E. coli. Consequently, identified sequences
can be prepared in amounts adequate for manipulation. Detailed
information about the organization of only a few very limited
regions of the chromosome will be available soon. But among the
greatest mysteries, the greatest uncertainties are techniques for
the introduction of new genetic material into the cell and for
insuring its integration into the replicating system of the plastid.

The relative simplicity of the plastid chromosome makes the
prospect of genetically engineering the cell through its plastid
genome seem tantalizingly attainable although it is still far off.
Prospects of genetically engineering plants through the plastid
genome either to solve basic research questions or for applied
purposes is especially appealing because of the difficulty of
selecting some desirable features in relatively slow-growing intact
higher plants. Part of the appeal and enthusiasm for this line of
work comes from the ability to regenerate sexually competent whole
plants from naked protoplasts in a number of species. The sight
of the thin plasma membrane at the outside of the protoplast seems
to make it an almost too vulnerable recipient for foreign DNA or
plastids and heightens the sense that DNA can be introduced and
that the organism's genetics can be modified at will. The impression
of imminent success is premature if not false.

This paper will begin with a review of plastid chromosome
organization and plastid differentiation followed by outlines of
some interesting problems which may be uniquely soluble by techniques
of genetic engineering. There are many unresolved problems which
are not necessarily insoluble or insurmountable but at this time we
have little basis for judging how far in the future or how difficult
the solutions may be. Genetic engineering of plants through their
plastid genomes will have a significant effect on basic research.
We have little basis for judging how soon or if genetic engineering
of plants through their plastid genomes will have a significant
effect on the production of current or new crops.

GENES FOR PLASTID COMPONENTS

The Genome of Zea mays Plastids

Covalently closed circular DNA molecules have now been identified
and in some cases isolated from chloroplasts of Euglena gracilis (5,
6), Antirrhinum majus, Oenothera hookeri, Beta vulgaris, Spinacia
oleracea (7), pea, bean, lettuce, oats and corn (8,9). Plastid DNA
(plDNA) molecules of higher plants are generally about 45 μm in
contour length. Their molecular weights range from 96.7×10^6 for
lettuce plDNA to 85.4×10^6 for corn plDNA. Molecular weights
based on renaturation kinetics agree with those estimated from
measurements of oontour length (8). Thirty-four percent of the

plDNA was obtained as 40 μm circles from Euglena chloroplasts (5). Up to 80% of the total DNA from higher plant chloroplasts has been found in circular molecules with 15 to 30% of the total as super-coiled circles. A small fraction of plDNA molecules occur as circular dimers.

On the average, a single chloroplast has 10 to 50 times the DNA in a single circular molecule (3,5,9-11). Renaturation kinetics indicated that there is a single type of plDNA in each chloroplast which is reiterated. This view is supported by data from the physical mapping of corn plDNA described below.

All of the recognition sites for the restriction endonuclease SalI and some of the recognition sites for endonucleases BamHI and EcoRI have been mapped on the maize chloroplast chromosome (12). This map is shown in Figure 1. Some portions of this chromosome have also been mapped (13) using endonucleases HindIII, BglI, PstI, and SmaI. The masses of all SalI-generated fragments of maize plDNA sum to 85 x 10⁶ daltons, close to the estimated molecular mass of the chromosome from electron microscopic measurements. This supports the renaturation kinetics data which indicate that there are numerous copies of a single type of chromosome in each chloro-plast. The physical mapping also reveals the presence of a 22,500

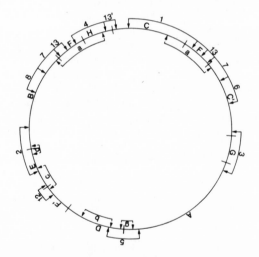

Figure 1. A physical map of the Zea mays plastid chromosome (12) showing all the recognition sites for the endonuclease SalI (frag-ments designated by capital letters, recognition sites by short straight lines across the circle); some recognition sites for EcoRI (fragments designated by lower case letters; recognition sites by arrow heads pointing outward from the inside of the circle); and some recognition sites for BamHI (fragments designated by Arabic numerals; recognition sites by arrows pointing inward from outside of the circle).

base pair long, singly repeated, inverted sequence in maize p1DNA.
This inverse sequence has also been observed in electron micrographs
(14). In the latter work, an inverted sequence 24,400 base pair
long was also found in chromosomes of spinach and lettuce chloroplasts. On the other hand, pea chloroplast DNA does not contain an
inverted repeat. The circular dimers of pea p1DNA occur in a head-
to-tail conformation whereas approximately 70 to 80% of the circular
dimers in preparations of lettuce and spinach p1DNA were found in
a head-to-head conformation. The latter were thought to be formed
between two circular monomers by recombination in the inverted
sequence. However, some circular dimers of spinach and lettuce
p1DNA were found in the head-to-tail conformation characteristic
of circular dimers of pea p1DNA which does not contain an inverted
repeat. The significance of inverted versus tandemly repeated sequences remains to be elucidated.

The statement that the sum of the SalI-generated fragments of
maize p1DNA is about 85×10^6 daltons must be qualified. All of
the fragments are present in unit amounts save one. That, designated
Sal G, is detected with a frequency of only 0.8. Either 20% of
p1DNA molecules lack Sal G or, equally likely at this time, this
fragment was lost during experimental manipulations.

Molecular hybridization studies with p1DNAs and p1rRNAs indicated that there are approximately two p1rRNA gene equivalents per
DNA molecule (15) from bean, lettuce, spinach, oats and maize.
There are three genes for each chloroplast rRNA species in the DNA
of E. gracilis chloroplasts (16).

In maize, 5S, 23S and 16S chloroplast rRNAs hybridize to p1DNA
EcoRI fragment a, a 12,000 base pair long segment present in each of
the two inverted repeats. This fragment has been cloned in E. coli
after incorporation into the vehicle plasmid pMB9. The genes for
these rRNAs have been mapped in detail by molecular hybridization
and electron microscopic observations of hybrids between cloned
EcoRI fragment a and rRNAs isolated from maize chloroplasts (17).
The orientation of these genes with respect to the smaller and
larger nonrepeated segments of maize p1DNA was also determined by
examination of R-loops remaining after hybridization of rRNA to
single-stranded p1DNA circles and DNA-DNA hybridization of segments
of the inverted repeat not hybridized to rRNAs. Conclusions from
these experiments (17) are shown in Figure 2.

The enzyme ribulose bisphosphate carboxylase (RuBPcase) often
comprises more than 50% of the soluble leaf protein. It catalyzes
the synthesis of two molecules of glycerate-3-phosphate from one
molecule of ribulose 1,5-bisphosphate plus one molecule of carbon
dioxide. The plant enzyme has a molecular weight of about 550,000
but it is comprised of two types of much smaller subunits. The
large subunit (LS) has a molecular weight of 51,000 to 58,000 while
the small subunit (SS) is 12,000 to 18,000 depending upon the species
(18). The gene for maize LS-RuBPcase has been located on the maize
p1DNA within a 2500 base pair long sequence (13,19). The location
of this gene is shown in Figure 2.

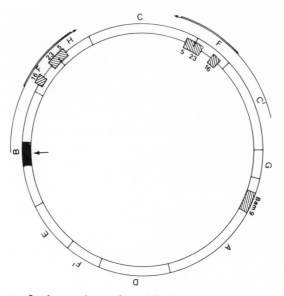

Figure 2. A map of the maize plastid chromosome showing (a) the
locations of recognition sites for the restriction endonuclease
SalI (12) (lines connecting the two concentric circles representing
two strands of DNA); (b) the locations on the inverted repeated DNA
sequences (arrows) of genes for 16, 23 and 5S rRNAs (17); (c) the
location of Bam fragment 9, which contains the structural gene for
the large subunit of RuBPcase (13,19); and (d) the location of Bam
fragment 8 (shown by the solid black zone within Sal fragment B and
indicated by an arrow) which contains the gene for a 32,000 dalton
thylakoid membrane protein (20).

    The 4200 base pair fragment (Bam 9) has been cloned in E. coli
using as a vehicle the plasmid RSF1030. The LS-RuBPcase gene was
mapped on the maize plastid chromosome by, on the one hand, establishing the site of Bam 9 within Sal fragment A, and on the other
hand, showing that the chimeric plasmid as well as isolated Bam 9
and certain subfragments of Bam 9 specify the synthesis of LS polypeptide in a linked transcription-translation system. The identity
of the polypeptide produced in vitro with LS obtained from maize
leaves was established immunochemically as well as by comparison of
proteolytic fragments (13,19).
    Another structural gene has been located on the maize plastid
chromosome. The gene for a 32,000 dalton photosynthetic membrane
(thylakoid) protein which is absent from etioplasts of dark-grown
(i.e., etiolated) seedlings is produced in large amounts in greening
seedlings and plastids isolated from them (20-22). The site of the
structural gene for this polypeptide which is formed from a 34,500
dalton precursor (21) is shown in Figure 2. A chloroplast DNA sequence
containing this gene has been inserted into pBR322 and cloned in
E. coli (L. McIntosh and L. Bogorad, unpublished data).

In addition to the genes for rRNAs, LS-RuBPcase and the 32,000 dalton thylakoid polypeptide which have been assigned positions on the maize plastid chromosome, hybridization to total plDNA shows that 0.60 to 0.75% of the chromosome consists of sequences complementary to maize tRNA, corresponding to 20 to 26 tRNA cistrons. It is known that tRNAs charging a total of at least 16 different amino acids hybridize with maize plDNA (23).

Saturation hybridization experiments indicate that virtually all of the maize plDNA is transcribed in chloroplasts (L.A. Haff and L. Bogorad, unpublished data). The same seems to be the case for tobacco (15) and E. gracilis (24). In maize, taking into account the approximately 15% of the genome which is repeated plus space occupied by genes for rRNAs and tRNAs, the remaining DNA sequences could be adequate to code for 85 to 100 polypeptides in the 25 to 30,000 dalton size range. This calculation assumes complete transcription of the equivalent of one strand of DNA and complete translation of transcripts. We do not know whether the assumptions are valid.

Genes for Chlamydomonas Chloroplast Ribosomal Proteins

By the techniques of transmission genetics Mendelian, bi-parentally inherited genes for chloroplast functions can be distinguished from nonMendelian, uniparentally inherited genes in the green alga Chlamydomonas reinhardi. All nonMendelian mutations in this alga have been traced to a single linkage group and several antibiotic resistance markers have been mapped. Various lines of indirect evidence indicate that the nonMendelian genome is in the chloroplast (25). A proposed arrangement of these loci shown in Figure 3 is taken from Boynton et al. (26). The effects of a few of these mutations are known. A mutation at ery-U-1 confers resistance to the antibiotic erythromycin and appears to alter protein LC4 of the large subunit of the Chlamydomonas chloroplast ribosome (27,28). A mutation to streptomycin resistance which maps to the locus sr-u-2-60 affects the assembly of the chloroplast ribosome's small subunit (29) probably as a result of a mutation in rRNA or a protein. A mutation to streptomycin resistance which maps to the locus sr-u-sm2 may alter a protein of the small subunit (30).

On the other hand, a mutation to erythromycin resistance which has been mapped to locus ery-M-1 on nuclear linkage group XI is in the structural gene for chloroplast ribosomal protein of the large subunit, LC6 (31). Another nuclear erythromycin-resistant mutant, ery-M-2d, has a different ribosomal large subunit protein that is altered (28).

Thylakoid polypeptide 5 of Chlamydomonas has an apparent molecular weight of about 50,000. A chloroplast gene mutant, thm-u-1, contains a variant polypeptide that is larger than the wild-type form by about 1000 daltons (32). This mutation has not been mapped but its genetic transmission pattern clearly shows it to be in the uniparental genome.

L. BOGORAD

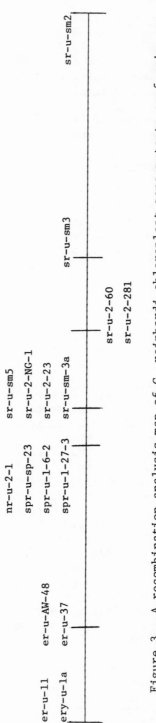
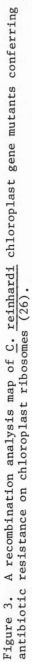

Figure 3. A recombination analysis map of *C. reinhardi* chloroplast gene mutants conferring antibiotic resistance on chloroplast ribosomes (26).

Other Genes for Chloroplast Components

Two approaches besides transmission and in vitro genetics have been employed to attempt to locate other genes for chloroplast components in the nuclear or plastid genomes. One line of experimentation is based on the idea that antibiotics such as lincomycin and chloroamphenicol block the action of organelle but not of cytoplasmic ribosomes while cycloheximide exclusively interferes with protein synthesis by cytoplasmic ribosomes. The chloroplast (or organelle) versus cytoplasmic ribosome dichotomy may not be far off the mark for in vitro analyses but in vivo complications with regard both to effects of cycloheximide on other reactions and membrane barriers reduce confidence that all the responses to these agents are understood and under control in the living cell. Another underlying assumption is that the compartment in which an mRNA is translated is the same in which the gene is transcribed. Proteins synthesized in the cytoplasm are viewed as products of nuclear genes and chloroplast-synthesized proteins as products of plastid genes. This is a reasonable working hypothesis but must be recognized for what it is--a convenient assumption, not a fact. Some of the most successful experiments of this sort (33-35) have been carried out in steps with microorganisms: yeast to study the production of mitochondrial proteins; C. reinhardi to seek the source of chloroplast proteins. First, for example, protein synthesis might be permitted to go on in the presence of chloramphenicol, then the cells are washed free of this antibiotic and incubated with cycloheximide. With the use of appropriately labeled amino acids it has been possible to determine whether a polypeptide is produced in the presence or absence of one inhibitor or the other yet the final cell has, presumably, normal completed components.

In another approach to the problem, chloroplasts are isolated and supplied with radioactive amino acids such as ^{14}C- or ^{3}H-leucine or ^{35}S-methionine to label the polypeptides produced with light as the source of energy for the photosynthetic ATP-generating system (22,36,37). The underlying assumption in these experiments is obvious. It is fundamentally the same as that made in the inhibitor experiments. Namely, messenger RNA is translated in the cellular compartment containing the gene from which it has been transcribed-- isolated chloroplasts translate messenger RNA transcribed from chloroplast genes.

GENE DISPERSAL AS A PRINCIPLE OF ORGANELLE BIOLOGY

Very few structural genes for identified RNAs or polypeptides have been located on the chromosome of any chloroplast and only one gene for a chloroplast component has been firmly located in the nuclear genome: C. reinhardi chloroplast ribosomal protein LC6 (31). Yet, a principle of gene distribution for multimeric organelle com-

ponents has begun to emerge. In <u>Chlamydomonas</u> <u>reinhardi</u>, genes for
some chloroplast ribosomal proteins are in the nuclear genome and
others in the chloroplast genome; the genes for chloroplast rRNAs
are situated in the plastid genome. The gene for the LS of RuBPcase
has been mapped physically in the maize chloroplast genome (19) and LS
is produced by isolated plastids (36) but in <u>Nicotiana</u> the gene for
the small subunit of RuBPcase is transmitted in a Mendelian
manner (38).

At least 13 polypeptides synthesized by isolated maize plastids
correspond in molecular weight to thylakoid components while a
number of other thylakoid proteins are not synthesized by isolated
plastids (22) and thus are believed to be nuclear genome products.
Among the thylakoid membrane (photosynthetic membrane) proteins
made by isolated maize plastids (32) is one of 32,000 daltons whose
gene has been mapped on the chloroplast genome (20) but polypeptides
of the light-harvesting chlorophyll-protein complex appear to be
coded by nuclear genes (based on the pattern of genetic transmission
in <u>Nicotiana</u>), are conspicuously absent from among the products of
isolated plastids (22), and are synthesized from poly A$^+$ RNA obtained
from the cytoplasmic fraction of barley (39).

Three of the five subunits of the thylakoid coupling factor
(CF$_1$) are synthesized by isolated spinach chloroplasts (37). Two
of these have been detected among the products of isolated maize
plastids (21). It seems likely that other subunits of CF$_1$ are produced in the cytoplasm from information in nuclear genes.

The phenomenon of gene dispersal (40,41) is important in the
present discussion for two reasons. First, genetic engineering
techniques are likely to be important for discovering mechanisms
of gene dispersal and the stabilization of genomes. Second, in
practical applications, control over the nuclear genome, for
example, may be exerted by altering an organelle genome or vice
versa. Evolutionary considerations leading to gene dispersal have
been discussed previously (40,41).

PLASTID PHYSIOLOGY AND BIOCHEMISTRY

The Chloroplast

The best known type of plastid is of course the chloroplast.
This organelle is specialized for the capture of light energy which
drives the photosynthetic machinery. Water is split, oxygen is
released and sugar is made from carbon dioxide. All of photosynthesis, save fixation of carbon dioxide into a carboxylic acid and
the reduction of the latter to carbohydrate, is carried on in the
specialized chlorophyll-bearing lamelli which form the thylakoids--
osmotically responsive sacs. In higher plants, the thylakoids are
generally arranged in stacks called grana. The membrane-free regions

in the chloroplast are called the stroma. Some lamellar elements
extend between grana. These, too, carry light-harvesting and energy
transducing apparatus whose products are ATP, NADPH and O_2. Electron
transparent regions in the stroma contain threads of chloroplast DNA.

Chloroplast Differentiation in C4 Plants

Plants of the type called C3 fix carbon in the photosynthetic
carbon reduction cycle. The enzyme ribulose bisphosphate carboxylase (RuBPcase) catalyzes the production of two molecules of phosphoglyceric acid from one molecule of ribulose 1,5-bisphosphate and
one molecule of carbon dioxide (18). The organic acid produced is
reduced to phosphoglyceraldehyde using ATP and NADPH produced by
the chloroplast thylakoids. Other plants, designated C4, work
differently. Chloroplasts in mesophyll (middle leaf) cells of these
plants lack RuBPcase. In these cells carbon dioxide is first fixed
into a 4 carbon acid (hence C4) by addition to a 3 carbon acceptor.
The 4 carbon acid, containing the newly added CO_2, is believed to
move to the chloroplast-containing cells which surround the vascular
bundles. In these bundle sheath cells, in some cases through the
cooperation of mitochondria, cytoplasm and chloroplasts, the 4
carbon acids arriving from mesophyll cells are decarboxylated and
the CO_2 is refixed, this time in a RuBPcase catalyzed reaction, into
phosphoglyceric acid. The bundle sheath cell chloroplasts have the
same RuBPcase carbon fixing machinery as chloroplasts of C3 plants.
Attractive features of C4 plants are their capacity for high photosynthetic rates because O_2 production at high light intensities
appears not to affect CO_2 fixation, they exhibit no or very low
photorespiration, and they lose markedly less water per unit of
dry matter produced than C3 plants.

PEP (phosphoenol pyruvate) carboxylase, the enzyme responsible
for fixing CO_2 to form the 4 carbon acid oxalacetate in mesophyll
cells of C4 plants, has a very high affinity for CO_2. C4 leaves
have very high stomatal resistances. This reduces the rate of water
loss but also results in lower levels of CO_2 in the intracellular
spaces of the leaf. Data on diffusion resistance and photosynthetic
rates for several species give an estimate of the concentration
of CO_2 in the stomatal liquid phase of C3 plants as about 6 µM compared with 1 to 2 µM for C4 plants. At these lower values the PEP
carboxylase reaction can proceed easily but RuBPcase activity would
be deficient by several fold to account for the maximum rates of
photosynthesis by C4 species. Thus, in a way, C4 species can afford
to decrease gas transfer with the environment and thus operate at
lower water loss rates (42).

RuBPcase can also operate as an oxygenase. Oxygen appears to
compete with CO_2. At high concentrations of CO_2, two molecules of
phosphoglyceric acid are produced for each carbon dioxide taken up.
In the absence of CO_2 but presence of oxygen, one molecule of phos-

phoglyceric acid and one of phosphoglycolate is formed per molecule
of ribulose 1,5-bisphosphate consumed. The relative oxygenase/
carboxylase activity depends upon the partial pressure of oxygen.
Of course, oxygen is a product of green plants' photosynthesis and
the higher the light intensity, the greater the rate of photosynthesis, the greater the production of O_2, the greater the inhibition
of CO_2 fixation by promoting the oxygenase reaction of RuBPcase.
The C4 mechanism provides the possibility of higher CO_2 concentrations in the bundle sheath cells thus favoring the carboxylase over
the oxygenase reaction of RuBPcase.

Photorespiration is the oxidation of phosphoglycolate. Photorespiration is "respiration" because oxygen is consumed and CO_2 is
evolved. It operates only in light because its substrate is phosphoglycolate produced together with phosphoglyceric acid in the
presence of O_2 by RuBPcase in the photosynthetic carbon reduction
cycle. C4 plants also contain lower activities of some of the
enzymes essential for C4 photorespiration (42).

Mesophyll cells of C4 leaves contain enzymes for C4 fixation
but lack RuBPcase. Bundle sheath cells of the same leaves contain
RuBPcase and enzymes for decarboxylating malate or aspartate. Both
mesophyll and bundle sheath chloroplasts of maize contain the gene
for the LS of RuBPcase. The mRNA transcript of this gene is un-
detectable in mesophyll cells though abundant in bundle sheath cells
(43). Sites of other genes for C4 enzymes are not known at the
present but this is an interesting problem in chloroplast differentiation; converting C3 to C4 plants is one of the most intriguing
candidate modifications for genetic engineering.

Chloroplasts, Chromoplasts, Amyloplasts, Eliaoplasts

Green tomatoes turn red as they ripen. The color change is a
manifestation of chloroplasts losing chlorophyll and accumulating
carotenoids. The red, carotenoid crystal-containing plastid is
called a chromoplast. Some plastids are specialized all or part
of their lives as sites of starch storage, as in the case of amyloplasts of potato tubers and other tissues. In some cases, as starch
accumulates the thylakoids are eliminated and as the starch is used
up thylakoids are regenerated (44). Other plastids, elaioplasts,
are specialized for oil storage (4). We do not know the mechanisms
regulating the conversion of chloroplasts to chromoplasts, amyloplasts to chloroplasts and vice versa, etc.

GENETIC ENGINEERING

Mechanics of Genome Manipulation

There are five bodies of knowledge and technical experience
required for reengineering a plant through its plastid genome.

First, the DNA sequences to be introduced, whether they be completely synthetic or from a foreign source such as a bacterium, blue-green alga, the nuclear or plastid genome of another plant species, or an animal genome (with or without synthesized additions) must be chemically and biochemically well characterized and abundantly available in pure form. Second, the organization of the recipient plastid genome should be understood at least well enough; for initial experiments, detailed information about some small segments containing an identified gene would suffice; for greater flexibility, detailed knowledge of more of the chromosome needs to be in hand together with an understanding of mechanisms of replication and regulation of gene expression. Third, a DNA sequence must be transported into the cell and into the plastid. Fourth, the foreign DNA must either be incorporated into the replicating system of the plastid genome or introduced as an independent replicating element. Fifth, the expression of the genetic material that is introduced must somehow be regulated or, if desired, completely deregulated.

The first problem, having well characterized sequences available in adequate amounts (i.e., cloned) is coming close to being resolved in a very limited way as cloning, DNA sequencing, techniques for assessing the information content of cloned sequences and initiation and terminating signals on DNA all develop. Most of the maize chloroplast genome is cloned in E. coli (13,17,19; J.R. Bedbrook and L. Bogorad, unpublished data; L. McIntosh and L. Bogorad, unpublished data) using pMB9, RSF1030, pML21, pBR322 and other vehicles and a few specific cloned genes are available for detailed analysis-- perhaps for introduction experiments. Isolation, cloning and characterization of DNA sequences from other sources for use in plant cell experiments is beyond the scope of this paper.

Then there is a second set of problems. How is the plastid genome organized; how is its replication and expression regulated? In general, what is likely to happen to an introduced fragment based on what happens in the unperturbed system? Physical maps of plastid chromosomes of a number of species are likely to be forthcoming in the next few years. Details of the organization of small regions of already explored genomes are likely to appear on about the same sort of timetable.

Plant cells can be enzymatically stripped of their walls of cellulose, hemicelluloses and pectins. Protoplasts can be manipulated and maintained in isotonic media but under the proper hormonal conditions begin to regenerate outer walls. Undifferentiated masses of callus tissue can be generated from the individual isolated plant cells and, in some cases, whole plants can be regenerated.

The most apparent and simplest way one can imagine for getting DNA into a plant cell would be to mix protoplasts with the DNA sequence of interest under conditions where plasma membranes are likely to fuse with other membranes. Polyethylene glycol promotes fusion and so do calcium ions at higher pH values. But first, how can DNA be protected from destruction by nucleases before it gets into the

cell? Nucleases need to be eliminated or inhibited by selection of
the proper conditions and medium. Alternatively, both to protect
the DNA and promote uptake, the DNA could be wrapped in a protein
(45,46), or other material which may fuse directly with the plasma
membrane or pass into the cell and be unwrapped enzymatically. In
either case, the DNA would be released into the cell after getting
there as a kind of artificial virus. The next problem is to get
the foreign DNA into the sphere of the genomic system in which it
will replicate and be transcribed. In the present case, how to get
it into the chloroplast? There is no strong evidence that nucleic
acids routinely pass across the plastid membrane. Yet it may be
occurring and the problem of getting the DNA in may not be very
serious. Anyone who has tried isolating etioplasts can attest that
their outer membranes are much more fragile than those of chloroplasts. DNA may pass across the membrane more easily at some
physiological stages or at some points in the cell cycle than others.
The recitation of possibilities reveals our ignorance of the problem
which is likely, in the end, to be solved empirically. Another line
of attack stems from the likelihood that at some time during the
evolution of eukaryotic cells genetic material was exchanged (40,41)
and rearranged among the cell's genomes. Mutants with unstable
nuclear-plastid genome relationships may have to be sought--however
those would be selected!

What about introducing DNA into plastids and the plastids into
cells? The successful functional introduction of DNA into plastids
cannot be tested at this time without having plastids survive and
multiply after being taken into cells because there are no reports
of plastids surviving in vitro for days or even hours while con-
tinuing their normal in vivo activities. Consequently, to test
whether DNA has been introduced into a plastid, the organelle must
be introduced into a protoplast which, in turn, must grow into a
callus and perhaps even a plant. Wild carrot (Daucus carota)
protoplasts in polyethylene glycol-containing medium take up chloroplasts prepared from the algae Codium fragile or Vaucheria (47).
The structurally distinctive chloroplasts of Vaucheria could be
identified in electron micrographs of carrot protoplasts. The long
term survival and the potential for replication of these transplanted
chloroplasts had not been investigated at the time the report was
published. As mentioned in an earlier section, those multimeric
components of organelles that have been studied to date are comprised of some elements coded for by nuclear genes and other elements
coded for by plastid genes. The requirement for gene product
matching probably imposes limits on the success or stability of combinations as phylogenetically diverse as Codium chloroplasts in
Daucus protoplasts but the possible range for organelle persistence
remains to be studied.

Kung et al. (48) reported that protoplasts prepared from white
portions of variegated Nicotiana tabacum took up green chloroplasts
prepared from N. suaveolens and green plants were formed. It now

appears more likely (P.S. Carlson, personal communication) that the green plants were the products of two fused cells. But, in principle, there seems to be little reason why plastids introduced into a cell cannot be maintained and replicate.

Suppose foreign DNA is introduced into a plastid directly or indirectly. How can its destruction, neglect or ejection be blocked? How can it be integrated into the plastid's replicative system? First, the DNA sequence of interest could be bordered with known chloroplast DNA sequences obtained by molecular cloning with the hope that the foreign DNA would be integrated into the chromosome through exchange between the complementary borders and regions in the plastid chromosome. A second alternative is to design a plasmid that would replicate in the plastid and to incorporate the DNA sequence of interest into it. Such a plasmid vehicle would consist of a replication starting point of chloroplast DNA and some selective marker. Replication starting points in chloroplast DNA have been identified in the electron microscope (49) but not physically. The construction of a chloroplast minichromosome should be possible after replication sites have been isolated.

Genetic Engineering to Solve Basic Research Problems

The presence of multiple, functionally integrated, compartmentalized genomes is a characteristic of eukaryotic cells. Understanding the biology of eukaryotic cells requires comprehension of principles of gene dispersal and intergenomic integration (40,41). Once developed, plastid genetic engineering techniques would provide methods for attacking many of these and other basic biological problems.

For example, by introducing a DNA sequence including a structural gene for a product distinguishable from components of the normal plastid together with its associated promoter, terminator, and other nontranscribed regions as part of a plastid minichromosome, it would be possible to study the effects of deletion of putative control or integrative elements of the introduced sequence on expression of the gene. Or, an extra copy of a normally occurring but distinctive, identifiable gene might be introduced either on a minichromosomal plasmid plastid or bounded by bordered sequences to study the control of expression of genes normally present in the chromosome. For some of these problems the number of copies of the chloroplast chromosome present per organelle may be significant and methods might need to be developed to reduce the number as a way of insuring--in the insertion mode--that all of the chromosomes are altered. A DNA sequence might be insertable into the center of the gene (by placing some nonsense sequence in the middle of bordering regions that are the sequences of a structural gene) to study the effect of deletion of this function on plastid development or function. Also in this way, a suspected control sequence outside of a structural gene might be tested by elimination.

As pointed out repeatedly above, multimeric components of organelles appear generally, perhaps universally, to be comprised of products of nuclear and organellar structural genes. Understanding the rules of gene dispersal may be a problem uniquely resolvable by genetic engineering techniques. We can imagine introducing, either via the insertion method or a minichromosome plasmid plastid, a gene normally found in the nucleus. Methods would have to be devised to determine whether the gene, obviously with some signature characteristic, is retained in the plastid for generations, lost rapidly, etc. Any structural elements that might alter this behavior could be explored using this technology.

Chloroplasts may contain 30 to 50 copies of the same chromosome (3,12) and leaf cells perhaps 40 chloroplasts (4). Each gene present once per plastid chromosome is in 1200 to 2000 copies per cell, as a rough example. Are structural genes for nuclear coded components of multimeric plastid elements such as some ribosomal proteins also present in 1200-2000 copies? Or, are plastid genes largely unused? Effects of numbers of some plastid genes--increasing the number by introducing minichromosomes or reducing the number of copies of some other genes by introducing destructive nonsense sequences-- might provide an experimental entree.

Genetic Manipulation for Crop Improvement

This section is written like science fiction and should be read that way. If none of the fantasies are realized, consider it to have been written for amusement. If some possibility only mentioned in passing is realized, the passage must be taken as perceptively prophetic--which it is! This sort of prognostication is saved because the record shows the most unbridled imagination to appear unorginal compared to eventually realized discoveries and innovations.

The chloroplast has evolved as an energy transducer. It has the great power of converting the energy of photons into chemical bond energy--most notably into a high energy phosphate bond in ATP. Many plastid genetic engineering applications can be considered first as rearrangements for diverting the light-generated chemical bond energy normally used for carbon dioxide fixation to other synthetic purposes such as nitrogen fixation or the production of industrially useful chemical compounds.

Our knowledge of fundamentals of energy transduction in photosynthesis is growing rapidly but it remains too meager to suggest modifications that would increase the efficiency of light absorption (short of altering light-harvesting systems to make them more like those in some blue-green or red algae); or would beneficially alter the electron transport or proton pumping systems; or would increase the efficiency of high energy phosphate bond synthesis;

or would convert to biologically useful forms the energy potential of the strong oxidant generated by the photolysis of water.

It is much more apparent how we might consider genetically reengineering C3 plants to the photosynthetic carbon metabolism patterns of C4 types. The problem is difficult, requires detailed analysis and can be approached but slowly. As outlined earlier, mesophyll cells in C4 plants are specialized for different enzymatic steps in carbon fixation than bundle sheath cells of the same leaf. Mesophyll plastids lack RuBPcase but contain pyruvate P_i dikinase, an enzyme that catalyzes the formation of phosphoenol pyruvate from pyruvate (imported from the cytoplasm) using photosynthetically generated ATP. The chloroplast-synthesized phosphoenol pyruvate passes into the cytoplasm where it is carboxylated to oxalacetate which is taken into the chloroplast where it is reduced to malate by NADPH produced photosynthetically. The malate (or the aspartate derived from it) is moved from mesophyll cells to bundle sheath cells where it is decarboxylated. The CO_2 released is fixed into phosphoglyceric acid by RuBPcase. In C4 plants of the NADP-ME type (e.g., Z. mays), decarboxylation of malate and production of phosphoglyceric acid both occur in the chloroplasts of bundle sheath cells; in other types of C4 plants, decarboxylation of a 4 carbon acid and CO_2 refixation involves mitochondrial and/or cytoplasmic as well as plastid enzymes of bundle sheath cells (42). Pyruvate, the other product of decarboxylation of malate, is transported back to the mesophyll cells where it can be used for the synthesis of additional phosphoenol pyruvate.

Pyruvate P_i dikinase has not been detected in C3 plants (except those of the crassulacean acid metabolism type). The activities of half a dozen additional enzymes are quantitatively different in C3 versus C4 plants. It is not known whether the genes for pyruvate P_i dikinase or any of the other enzymes are in the nuclear, mitochondrial or chloroplast genome.

It seems very complex to consider engineering into a plant all of these differences in cell type, etc. However, it has been known since early in the study of extranuclear inheritance that plastids of two different types may coexist in a single cell (4). Perhaps C4 carbon metabolism can be engineered into C3 plants by having both mesophyll-type and NADP-ME bundle sheath-type plastids in the same cell? The NADP-ME bundle sheath-type plastid is very similar to the C3 plastid with the most conspicuous exception being a 10-fold or so higher activity of NADP-malic enzyme (42). This quantitative difference is most likely to be a gene regulation problem in the plastid or nuclear genome. Perhaps a change is required in an otherwise normal plastid of a C3 plant. This modification is one in a long list required for the engineering surgery. It is one of many for which we lack basic knowledge.

What alterations are needed in a C3 plastid to produce a mesophyll-type of C4 chloroplast? The gene for LS-RuBPcase needs to be removed or functionally inactivated by one or another of the

techniques mentioned in an earlier section, or by the more formidable
task of eliminating all of the plDNA in some plastids and introducing
a remodeled chromosome lacking this gene, then introducing such a
plastid into a protoplast still containing some normal type plastids
but with increased NADP-malic enzyme. Other problems to be faced are
how to increase by 10- to 100-fold the activity of the cytoplasmic
enzyme phosphoenol pyruvate carboxylase, by 30 to 50 times the activity of adenylate kinase and by 10 to 30 times the activity of pyro-
phosphatase--from the levels in C3 cells to those in mesophyll cells
of C4 species.

There are two underlying problems in engineering an approach
that depends upon maintaining two chloroplast types in a single
cell. First is the assumption that genetic recombination does not
occur among plastids in a single cell of a higher plant. There is
some precedence for this in the persistence of two distinct plastid
types in single cells but studies of such plants reveal another
problem with the general approach. Sometimes the two plastid types
present in single cells segregate into separate tissue sectors. If
this happens within a leaf, it would only imitate more closely the
mesophyll-bundle sheath separation now observed in C4 plants. If
such segregation took a more usual pattern, there might be whole
branches with mesophyll-type plastids and others with only bundle
sheath-type plastids but probably only the latter would survive.
The outcome would be some plants with mixed plastid types and others
reverting essentially to C3 photosynthetic characteristics.

The mere recitation of the problems to be overcome reveals how
great the obstacles are. Yet, the outcome of experiments directed
toward making these modifications would reveal a great deal about
plastid-plastid and plastid-nucleo-cytoplasmic relationships.
Yield of the applied research is more problematical. If the plan
outlined in the few paragraphs above worked, RuBPcase would be
present only in those plastids with a high capacity for decarboxylating malate. These chloroplasts would have relatively high CO_2
to O_2 ratios internally which should minimize photorespiration. A
major advantage of C4 carbon metabolism is the ability to operate
with stomata having high diffusion resistances. As discussed in
an earlier section, this permits plants to grow with lower rates
of water loss. How can stomata be modified without knowing about
guard cell-epidermal genetics? Perhaps artificially applied anti-
transpirants might be used on C3 plants reengineered with C4 photosynthetic carbon metabolism.

Another attractive objective (or fantasy) for those contemplating
careers in plant genetic engineering is the conversion of nonlegumes--
especially cereals--into nitrogen fixing plants. The enzyme nitrogenase is produced in nodules of leguminous plants within the bacteroids formed from invading Rhizobium bacteria. Some cortical cells
of the root proliferate, enlarge and produce large amounts of leghemoglobin in response to infection by this bacterium. The hemo-
protein facilitates the transport of the large quantities of oxygen

required for oxidative phosphorylation by the bacteroids. Enormous amounts of ATP are necessary for nitrogen fixation. The leghemoglobin simultaneously keeps oxygen tensions relatively low around the nitrogenase which is sensitive to O_2.

Plastids normally contain nitrite reductase (49,50), glutamine synthetase (51) and ferrodoxin(52). To be capable of fixing nitrogen, plastids would need to have the gene for nitrogenase added, the capacity for O_2 production eliminated (because O_2 destroys nitrogenase) and perhaps some new electron transport components introduced. In line with the fantasy built upon fantasy that characterizes this section, introducing a gene for nitrogenase from a blue-green alga, a photosynthetic bacterium or a free-living bacterium sounds like the simplest of all the tasks ahead; never mind that we do not now even know all the problems! Fanciful, though probably realistic, ways to do this have been described in an earlier section. Elimination of photosystem II activity seems a bit harder. Yet, elimination of photosystem II is exactly what happens when vegetative cells of blue-green algae are converted to heterocysts (53). Photosystem II may be destroyed by a loss of manganese with the consequent inactivation of the oxygen liberating enzyme (55). The changes that occur during cellular differentiation in filamentous blue-green algae have not been catalogued entirely (let alone understood), but these cells also lose the capacities to produce RuBPcase (54) by some unidentified mechanism. Research into heterocyst differentiation may lead not only to a source of nitrogenase genes for cloning and transplantation but also to learning how to convert chloroplasts to nitroplasts in their other features. In addition to the uncertainties regarding the stability of mixed plastid cells already enumerated, we must ask here whether oxygen production by some plastids within a cell would interfere with nitrogenase in other plastids. Vegetative blue-green algal cells exist next to heterocysts without interfering with nitrogen fixation by the latter.

It is obvious that once we learn to introduce genes for polypeptides like nitrogenase into plastids, we should also be able to introduce genes for polypeptides of high commercial value in medicine or agriculture. Architypical is growing insulin-bearing plants in the field or culturing photosynthetic cells that produce insulin or other desirable polypeptides in continuous flow, chemical factory-like apparatus.

The differentiation of plastids into chromoplasts, i.e., conversion of plastids largely to the synthesis of carotenoids, suggests the possibility of including in each cell plastids that would produce large amounts of isoprenoids or other organic compounds which might be used as feedstocks in the chemical process industry. (Nielsen et al. (56) have suggested that some plant crops might serve as sources of fuel and hydrocarbon-like materials.) Here, too, two types of plastids might be designed to coexist in a single cell but the possibility of plastids with higher than normal amounts of carotenoids (or their isoprenoid precursors) can also be visualized. Such modified plants might be grown in the field and harvested or cells might be grown in culture in a continuous production line using solar energy. The genetic engineering required is mostly for altering gene expression rather than introducing any new structural genes.

The prospects for applied utilization of genetic engineering principles are all highly attractive but the primitive state of our knowledge and technology emphasizes the large amount of work that remains to be done. We probably have the least feeling for problems of plastid biology and relationships of plastids to the nucleo-cytoplasmic system. We do not understand how the number of plastids per cell is regulated. We do not know whether all types of plastids in a mixed population in a single cell will multiply at the same rate; if they do not, the possibility of segregation into separate tissue sectors is greater. Even at the level of DNA manipulation, where optimism is high and self-confidence perhaps unreasonable, we know little about the control of the gene expression and particularly how to keep introduced genes in place and how to regulate their expression. The task is tremendous but the number of research workers directing their attention to these problems of plastid genetics and biology is minute.

SUMMARY

A chromosome of a higher plant plastid is a circle of about 1×10^8 daltons. Virtually all segments of the plastid chromosome of _Zea_ _mays_ have been coupled with plasmid vehicles and cloned in _E._ _coli_. Thus there appears to be no obstacle to obtaining large amounts of these DNA sequences for manipulation. Much work remains to be done to understand the detailed organization of individual plastid genes. Such work should reveal important biological principles as well as provide information about possible sites for insertion of foreign DNA.

Vehicles for inserting DNA into plastid chromosomes or which can persist as independent replicating elements with a plastid's genome can be imagined but have not yet been constructed. Neither have methods been developed for introduction of foreign DNA into plastids, either directly or indirectly.

The possibilities for manipulating the plant cell through its plastid genome for knowledge or profit are easily imagined. Practice still requires a great deal of research. Unfortunately, the number of people currently concerned with these problems is very small.

Acknowledgments: The preparation of this paper and various parts of the research from the author's laboratory result from support in part by the Maria Moors Cabot Foundation of Harvard University and research grants from the National Institute of General Medical Sciences, the National Science Foundation and the United States Department of Agriculture's Competitive Grants Program.

REFERENCES

1   Correns, C. (1909) Z. Vererbungsl. 1, 291-329.
2   Bauer, E. (1909) Z. Vererbungsl. 1, 330-351.
3   Woodcock, C.L.F. and Bogorad, L. (1971) in Structure and Function of Chloroplasts (Gibbs, M., ed.), pp. 89-128, Springer-Verlag, Heidelberg.
4   Kirk, J.T.O. and Tilney-Bassett, R.A.E. (1967) The Plastids, W.H. Freeman and Company, Ltd., London.
5   Manning, J.E., Wolstenholme, D.R., Ryan, R.S., Hunter, J.A. and Richards, O.C. (1971) Proc. Nat. Acad. Sci. U.S.A. 68, 1169-1173.
6   Manning, J.R. and Richards, O.C. (1972) Biochim. Biophys. Acta 259, 285-296.
7   Herrmann, R.G., Bohnert, H.-J., Kowallik, K.D. and Schmitt, J.M. (1975) Biochim. Biophys. Acta 378, 305-317.
8   Kolodner, R. and Tewari, K.K. (1975) Biochim. Biophys. Acta 402, 372-390.
9   Kolodner, R. and Tewari, K.K. (1972) J. Biol. Chem. 247, 6355-6364.
10 Kung, S.-d. (1977) Ann. Rev. Plant Physiol. 28, 401-437.
11 Howell, S.H. and Walker, L.L. (1976) Biochim. Biophys. Acta 418, 249-256.
12 Bedbrook, J.R. and Bogorad, L. (1976) Proc. Nat. Acad. Sci. U.S.A. 73, 4309-4313,
13 Bedbrook, J.R., Coen, D.M., Beaton, A.R., Bogorad, L. and Rich, A. (1978) J. Biol. Chem. (in press).
14 Kolodner, R. and Tewari, K.K. (1978) Proc. Nat. Acad. Sci. U.S.A. (in press).
15 Tewari, K.K., Kolodner, R., Chu, N.M. and Meeker, R. (1977) in Nucleic Acids and Protein Synthesis in Plants (Bogorad, L. and Weil, J.H., eds.), pp. 15-36, Plenum Press, New York, NY.
16 Stutz, E. and Vandray, J.P. (1971) FEBS Lett. 17, 277-280.
17 Bedbrook, J.R., Kolodner, R. and Bogorad, L. (1977) Cell 11, 739-750.
18 Jensen, R.G. and Bahr, J.G. (1977) Ann. Rev. Plant Physiol. 28, 379-400.
19 Coen, D.M., Bedbrook, J.R., Bogorad, L. and Rich, A. (1977) Proc. Nat. Acad. Sci. U.S.A. 74, 5487-5491.
20 Bedbrook, J.R., Link, G.L., Coen, D.M., Bogorad, L. and Rich, A. (1978) Proc. Nat. Acad. Sci. U.S.A. 75, 3060-3064.
21 Grenbanier, A.E., Coen, D.M., Rich, A. and Bogorad, L. (1978) J. Cell Biol. 78, 734-746.
22 Grebanier, A.E., Steinback, K. and Bogorad, L. (1979) Plant Physiol. (in press).
23 Haff, L.A. and Bogorad, L. (1976) Biochemistry 15, 4105-4109.
24 Rawson, J.R.Y. and Boerma, C.L. (1976) Biochemistry 15, 588-592.

25   Sager, R. (1972) Cytoplasmic Genes and Organelles, Academic
     Press, New York, NY.
26   Boynton, J.E., Gillham, N.W., Harris, E.H., Tingle, C.L.,
     Van Winkle-Swift, K. and Adams, G.M. (1976) in Genetics and
     Biogenesis of Chloroplasts and Mitochondria (Bucher, T.H.,
     Neupert, W. and Werner, S., eds.), pp. 312-322, North Holland
     Press, Amsterdam.
27   Mets, L. and Bogorad, L. (1971) Science 174, 707-709.
28   Mets, L. and Bogorad, L. (1972) Proc. Nat. Acad. Sci. U.S.A.
     69, 3779-3783.
29   Conde, M.F., Boynton, J.E., Gillham, N.W., Harris, E.H.,
     Tingle, C.L. and Want, W.L. (1975) Mol. Gen. Genet. 140,
     183-220.
30   Ohta, N., Inouye, M. and Sager, R. (1975) J. Biol. Chem.
     250, 3655-3659.
31   Davidson, J.N., Hanson, M.R. and Bogorad, L. (1974) Mol. Gen.
     Genet. 132, 119-129.
32   Chua, N.-H. (1976) in Genetics and Biogenesis of Chloroplasts
     and Mitochondria (Bucher, T.H., Neupert, W. and Werner, S.,
     eds.), pp. 323-330, North Holland Press, Amsterdam.
33   Mason, T.L. and Scahtz, G. (1973) J. Biol. Chem. 248, 1355-
     1360.
34   Tzagoloff, A. and Meagher, P. (1972) J. Biol. Chem. 247,
     594-603.
35   Ohad, I. (1977) in International Cell Biology 1976-1977
     (Brinkley, B.R. and Porter, K.R., eds.), pp. 193-203, The
     Rockefeller University Press, New York, NY.
36   Ellis, R.J. (1977) in Nucleic Acids and Protein Synthesis in
     Plants (Bogorad, L. and Weil, J.H., eds.), pp. 195-212,
     Plenum Press, New York, NY.
37   Mendiola-Morgenthaler, L.R., Morgenthaler, J.J. and Price, C.A.
     (1976) FEBS Lett. 62, 96-100.
38   Bourque, D.P. and Wildman, S.G. (1973) Biochem. Biophys. Res.
     Commun. 50, 532-537.
39   Apel, K. and Kloppstech, K. (1978) Eur. J. Biochem. 85, 581-
     588.
40   Bogorad, L. (1975) Science 188, 891-898.
41   Bogorad, L., Bedbrook, J.R., Davidson, J.N., Hanson, M.R. and
     Kolodner, R. (1977) in Genetic Interaction and Gene Transfer,
     Brookhaven Symposia in Biology: No. 29, pp. 1-15, Brookhaven
     National Laboratory, Upton, NY.
42   Hatch, M.D. (1976) in Plant Biochemistry (Bonner, J. and
     Varner, J.E., eds.), 3rd Ed., pp. 797-844, Academic Press,
     New York, NY.
43   Link, G., Coen, D.M. and Bogorad, L. (1978) Cell (in press).
44   von Wettstein, D. (1958) in The Photochemical Apparatus, Its
     Structure and Function, Brookhaven Symposia in Biology: No. 11,
     pp. 138-159, Brookhaven National Laboratory, Upton, NY.
45   Collins, J. and Hohn, B. (1978) Proc. Nat. Acad. Sci. U.S.A.
     75, 4242-4246.

46   Cassells, A.C. (1978) Nature 275, 760.
47   Bonnett, J.P. and Banks, M.S. (1977) in International Cell
     Biology 1976-1977 (Brinkley, B.R. and Porter, K.R., eds.),
     pp. 225-231, The Rockefeller University Press, New York, NY.
48   Kung, S.D., Gray, J.C., Wildman, S.G. and Carlson, P.S.
     (1975) Science 187, 353-355.
49   Ritenour, G.L., Joy, K.W., Bunning, J. and Hageman, R.H.
     (1967) Plant Physiol. 57, 881-885.
50   Neyra, C.A. and Hageman, R.H. (1978) Plant Physiol. 62,
     618-621.
51   O'Neal, D. and Joy, K.W. (1973) Nature New Biol. 246, 61-62.
52   Kok, B. (1976) in Plant Biochemistry (Bonner, J. and Varner,
     J.E., eds.), 3rd Ed., pp. 845-885, Academic Press, New York,
     NY.
53   Donze, M., Haveman, J. and Schierck, P. (1972) Biochim.
     Biophys. Acta 256, 157-161.
54   Winkenbach, F. and Wolk, C.P. (1973) Plant Physiol. 52, 480-
     483.
55   Tel-Or, E. and Stewart, W.D.P. (1975) Nature 258, 715-716.
56   Nielsen, P.E., Nishimura, H., Otvos, J.W. and Calvin, M.
     (1977) Science 198, 942-944.

MITOCHONDRIAL DNA OF HIGHER PLANTS AND GENETIC ENGINEERING

C. S. Levings III

Department of Genetics
North Carolina State University
Raleigh, North Carolina 27607

and

D. R. Pring

Science and Education Administration
United States Department of Agriculture
Department of Plant Pathology, University of Florida
Gainesville, Florida 32610

INTRODUCTION

The advent and development of techniques of recombinant DNA and genetic engineering in prokaryotes has led to much speculation concerning the application of this technology to the broad area of higher plant improvement. Central to the development of this technology is the acquisition of a suitable vector or vehicle molecule which could be used to transfer desirable genetic information into plant cells. Prominent potential vehicles include the Ti-plasmids of Agrobacterium tumefaciens, the DNA plant viruses such as cauliflower mosaic virus, and the chloroplast and mitochondrial genomes of higher plants.

An idealized vehicle for recombinant DNA studies of higher plants should have characteristics of survival during uptake processes in plant cells, be free of pathogenic effects in the recipient cell, be capable of stable chromosomal insertion, and be capable of carrying desirable, expressible, genetic determinants into cells. While it seems unlikely that this idealized vehicle is presently available, several characteristics of the S-plasmids or fertility elements of the S male-sterile maize cytoplasm are particularly worthy of further consideration.

It is the purpose of this review to discuss contemporary information relative to the mitochondrial genomes of higher plants, with special reference to maize (Zea mays L.), and the unique S-plasmids of the S male-sterile cytoplasm. The elucidation of the nature and function of plant mitochondrial DNA molecules is paramount to the evaluation of potential recombinant DNA vehicles for genetic engineering in higher plants.

ORGANIZATION OF MITOCHONDRIAL DNA IN HIGHER PLANTS

It is becoming increasingly apparent that the mitochondrial genome of higher plants may be the largest known of any life form. The variability of mitochondrial DNA (mtDNA) ranges from an approximate 10×10^6 in animals (1) to between 90 and 200×10^6 in potato, Virginia creeper, wheat, maize, sorghum and soybean (2-5).

As opposed to evidence that mtDNA molecules are identical in many organisms (1), studies of higher plant mtDNA suggest substantial heterogeneity, since molecules large enough to account for the high molecular weight have not been isolated. Electron microscopy of pea mtDNA suggested circular molecules of 30 μm contour length (6); 28 μm linear molecules were observed in potato mitochondria preparations (7). Renaturation kinetics suggested molecular weights of 66 to 77×10^6 for pea and 100×10^6 for potato; the latter value was suggestive of a molecular weight in excess of that obtained by contour length measurements. Although restriction endonuclease analysis of pea mtDNA has not been reported, potato mtDNA molecular weight values of 90×10^6 were observed upon restriction (2). High molecular weight values have been observed for each higher plant mtDNA which has been analyzed by restriction endonuclease fragment analysis.

The apparent discrepancy between the molecular weight values obtained by contour length measurement and values generated by restriction analysis could be rationalized by intermolecular heterogeneity of higher plant mtDNA. Evidence supportive of heterogeneity has been obtained from maize and soybean. In maize (8,9) three discrete classes of molecules were observed with contour lengths of 16, 22 and 30 μm. These molecules, if assumed to occur in equimolar ratio, could account for the minimum estimated molecular weight of maize mtDNA based on restriction analysis (116 to 131×10^6) (3). In soybean, a more complex distribution of contour length was observed, with seven classes ranging from 5.9 μm to 29.9 μm (10). Restriction analyses of soybean mtDNA resulted in molecular weight values in the range of 150×10^6 (5). A multimodal length distribution has also been observed for Virginia creeper mtDNA (Quetier and Vedel, personal communication), and restriction analyses suggest a molecular weight of 165×10^6 (2). Digestion of sorghum mtDNA by several restriction endonucleases resulted in values ranging from 100 to 200×10^6 (4).

   The possibility that intermolecular heterogeneity exists in
higher plant mtDNA, as noted by Shah et al. (8) for maize and
postulated by Quetier and Vedel (2) for higher plants in general,
seems compatible with the increasing data suggestive of high molecular weights based on restriction analysis. Intermolecular heterogeneity could result from a series of separate and distinct DNA
molecules within a mitochondrion or the presence of several kinds
of mitochondria within a cell.
   It is not known if the large numbers of restriction fragments
of higher plant mtDNA each represent separate and unique base
sequences, or if rearrangements result in the apparent high molecular weight. Heterogeneity of herpes simplex virus DNA (11) was
sufficient to generate a restriction molecular weight of 160 x 10^6,
even though the intact DNA molecule has a molecular weight of 95 to
100 x 10^6. Four distinct structural forms of this DNA were apparent,
differing in orientation of large subregions; each subregion was
characterized by inverted terminally redundant repeats. Thus the
apparently high genomic complexity resulted from intermolecular
heterogeneity, and did not indicate genetically distinct DNAs.
   To date the apparent heterogeneity of mtDNA of higher plants
has not been observed in lower plants, such as Neurospora,
Saccharomyces, and Aspergillus. Electron microscopy and restriction analysis of Neurospora crassa mtDNA suggest a 20 μm molecule,
with restriction fragments totaling 40 to 41 x 10^6 (12,13). Similar
data were obtained from Saccharomyces mtDNA, where the total molecular weight of restriction fragments and the genomic size is about
50 x 10^6 (14-16). The mtDNA of Aspergillus nidulans has been shown
to total only 21 x 10^6 by restriction analysis, a genetic complexity
expected from the contour length of the molecule (17). In each
case the restriction patterns map circularly, providing unequivocal
evidence of the continuous nature of the genome.
   In Virginia creeper, cucumber, wheat and potatoes, Quetier
and Vedel (2) have reported isolating mtDNAs as covalently closed
circular molecules from the lower band of dye-CsCl gradient. Our
own experiences have been somewhat different. We have observed,
by electron microscopic examination, covalently closed circular mtDNA
molecules from maize, soybean, and flax (9,10, Lockhart and Levings,
unpublished data). However, in the case of maize (Figure 1), a
lower band of a dye-CsCl gradient was not visible and the supercoiled
molecules were isolated by removing a fraction from the position
on the dye-CsCl gradient where the lower band was expected. On the
other hand, the mtDNA of flax is readily isolated from a visible
lower band of dye-CsCl gradients as covalently closed circular
molecules. Based upon these few studies, it appears that the native
configuration of mtDNA from higher plants is that of a covalently
closed circular molecule.

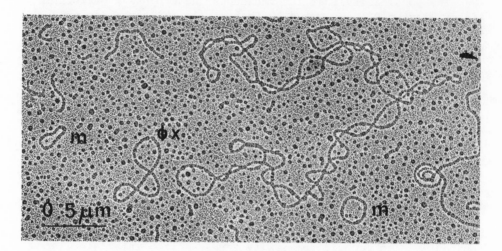

Figure 1.   Electron micrograph of a covalently closed circular
molecule of maize mitochondrial DNA; length is 14.6 μm.   Also
shown are two DNA minicircles (m); the internal standard, φX174
DNA (φX); and several DNA fragments.

Diversity of mtDNA Among Taxons

    Restriction endonuclease digestion of mtDNAs has been extremely
effective in identifying differences among mtDNAs derived from
various sources.   The limited numbers of endonucleases employed in
these studies suggest that substantial mtDNA diversity will continue
to be identified among mtDNAs of higher plants.   It is furthermore
apparent that the extent of this diversity reflects more than a
simple point mutation or limited base changes.   For example, the
mtDNA of the T source of cytoplasmic male sterility in maize is
clearly distinguished from that of normal, fertile maize by the
endonucleases HindIII, SalI, XhoI, EcoRI, SmaI, BamHI, and HaeIII
(3,18, unpublished data).   If the genetic complexity of maize mtDNA
is approximately 225,000 base pairs and if a restriction endonuclease requires six base pairs for recognition and produces 40
restriction fragments, only about 0.1% of the base sequences are
reflected in the resultant fragment pattern.   It seems surprising
then, that endonucleases should so clearly distinguish mtDNAs
within a single species.   Some of the normal, fertile cytoplasms
of maize can also be distinguished; five of nine cytoplasms examined
were separable by HindIII digestions of mtDNAs (19).   It seems probable that additional enzymes will differentiate within HindIII groups.
    Current investigations of mtDNA from fungi also suggest sub-
stantial mtDNA diversity among taxons.   Saccharomyces cerevisiae
mtDNA can be differentiated from that of S. carlsbergensis, and

strains of each species can also be separated (20). Substantial
progress has been made in identifying the nature of additions and
deletions which contribute to these variable restriction patterns
(15,16).

Comparison of Mitochondrial DNAs Within the Genus Zea

We have recently examined a series of cytoplasms of annual
teosinte (Zea mexicana) and perennial teosinte (Zea perennis).
The latter cytoplasm (often designated EP, or Euchleana perennis)
is associated with cytoplasmic male sterility in maize (21). When
mtDNAs of perennial teosinte and maize were compared, it was evident that the mitochondrial genome of Z. perennis was strikingly
different from that of Z. mays (22,23). The mtDNA from the annual
teosintes, Chalco, Guererro, Central Plateau, Heuheutenango,
Balsas and Guatemala, were subsequently compared to Z. perennis
and common Z. mays (24). Several restriction endonucleases were
utilized, and three groups of mtDNAs were observed: a) Z. perennis,
b) Guatemala Z. mexicana, and c) all other annual teosintes. The
difference between Z. perennis and Guatemala Z. mexicana was slight,
but the remaining annual teosintes were markedly different from
perennial and Guatemala teosinte. Interestingly, the five annual
teosintes resembled, in a general fashion, some of the Z. mays mtDNA
patterns previously obtained (3,19). Examination of the ctDNA of
these same taxa (24) did not result in the same alignments. Again,
three groups were observed: a) perennial teosinte and Guatemala
Z. mexicana, b) Huehuetenango and Balsas Z. mexicana, and c) Central
Plateau, Chalco, and Guerrero Z. mexicana. It is evident from these
studies that the two major organelles may have evolved separately,
since a) cases are apparent where two races share an apparently
common mtDNA in the presence of dissimilar ctDNAs, and b) cases
are apparent where two races share an apparently common ctDNA in
the presence of dissimilar mtDNA. Teosinte ctDNA displayed restriction patterns which were decidedly similar to those of Z. mays;
Central Plateau, Chalco, and Guererro Z. mexicana yielded ctDNA
patterns indistinguishable from those of many Z. mays inbred lines.

Collectively, these studies of members of the genus Zea have
suggested that there is substantial mtDNA heterogeneity among members of the genus and within the species Z. mays (3,19,22-24). At
least 14 different mtDNAs have been identified among all Zea
species examined to date. Since we have observed strict maternal
inheritance of mt and ctDNA in every example studied, we reason
that evolutionary divergence of these genomes is not of recent origin.

It is important to note, however, that the general patterns
within Zea mtDNA are similar; at least 75 to 80% of all fragments
appear to be of the same molecular weight. When Zea mays mtDNA
restriction patterns are compared with those from Sorghum vulgare,
very few fragments seem of identical molecular weight, and the

general patterns bear no resemblence to each other (Figure 2).
The fragment patterns would then seem to be of value in discerning
possible relationships among taxa.

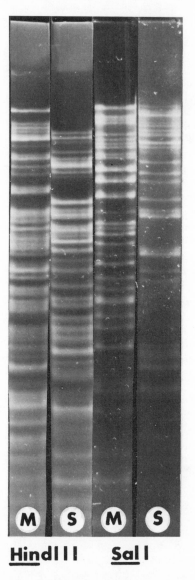

Figure 2. Agarose gel electrophoresis patterns of HindIII and
SalI digests of mitochondrial DNA from maize (M) and sorghum (S).

Informational Content of the Mitochondrial Genome of Higher Plants

Although in animal and yeast the informational content of the mitochondrial genome is largely known, very little is understood about the genetic content of higher plant mitochondrial genomes. It is likely that certain information encoded by the mtDNAs of these organisms will also be coded by plant mtDNAs. For example, certain polypeptide subunits of the proteins, cytochrome oxidase, cytochrome b, and ATPase are coded by animal and yeast mtDNAs, but this has not yet been verified with plant mtDNAs (1).

Higher plant mitochondria contain unique rRNAs which are distinct from cytoplasmic and chloroplast rRNAs (25,26). However, it has not been unambiguously established that these RNAs are encoded by higher plant mtDNA. Likewise, it has been demonstrated that higher plant mitochondria possess an unique 5S RNA, probably associated with the heavy rRNA (27,28). Generally, it is assumed that these mitochondrial RNAs are gene products of the mtDNA.

Although no functional assignment has been confirmed, discrete products of in vitro protein synthesis by isolated mitochondria have been identified (29). Significantly, the number and size range of these proteins were analogous to polypeptides thought to be gene products of mammalian and fungal mtDNA. It is evident from the brevity of this treatment that much remains to be learned about the informational content of mtDNA from higher plants.

The larger amount of mtDNA found in higher plants raises a question about what additional information is coded by the extra mtDNA. Several traits which are unique to higher plants are inherited in an extrachromosomal manner; cytoplasmic male sterility, disease susceptibility and reduced kernel size are a few examples of this sort. It may be that the additional mtDNA found in higher plants is involved in coding for these unique traits. Other extrachromosomally inherited traits that involve the photosynthetic mechanism are present in plants; however, they would seem the responsibility of the chloroplast genome. Finally, it is possible that some of the extra mtDNA in plants codes for no additional information, but instead serves only a spacer function. Indeed, AT-rich stretches have been described in yeast mtDNA which presumably serve a spacer function (30).

Cytoplasmic male sterility (cms), an extrachromosomally inherited trait, is widespread in the plant kingdom. It has been reported in 80 species and 25 genera (31). In maize, the tassel of the cytoplasmic male-sterile plant does not exsert anthers, and consequently, no pollen is shed. A less drastic form sometimes occurs in which deformed anthers are exserted but they contain only aborted pollen grains. The cms trait does not affect female fertility. Although a large number of cms types have been reported in maize, only a few, T, S and C, have been authenticated as unique (32). The various cms are distinguished on the basis of the nuclear genes that restore pollen fertility (20,32,33).

For example, the S cytoplasm is restored to pollen fertility by a single locus, Rf3, which is located on chromosome 2. Conversely, the cms-T is not restored by the Rf3 locus, but instead restoration requires the Rf1 and Rf2 loci on chromosomes 3 and 9, respectively. Differences in the mode of fertility restoration, disease suscep- tibility, and mitochondrial DNA restriction patterns also serve to distinguish the various sterile cytoplasms (34). Accumulated evi- dence now suggests that the cms trait is coded by the mitochondrial genome. Recently, a complete review of this subject was prepared (34). Therefore, only a brief treatment is presented here.

Restriction endonuclease fragment analyses have been carried out on the mtDNAs from maize with normal or fertile (N) and T, C, S cms cytoplasms (3,18,24). To date, seven different restriction enzymes have been employed in these analyses: HindIII, BamHI, SalI, XhoI, EcoRI, SmaI and HaeII. Specific cleavage sites are different for each of these endonucleases. The fragment patterns from the four cytoplasms, N, T, C and S, were readily distinguishable re- gardless of the restriction enzyme used. Furthermore, these dis- tinctions were demonstrated to be constant irrespective of the nuclear background. These results indicate that the four cyto- plasms, N, T, C and S, each contain their own, novel mtDNAs.

Since the chloroplast DNA (ctDNA) is a possible site for the cms trait, restriction endonuclease fragment analyses have been made on ctDNA from maize with N, T, C and S cytoplasms. These analyses showed that the restriction patterns from ctDNA from N, T and C cyto- plasm were similar with HindIII, SalI or EcoRI digestions. However, ctDNA from the S cytoplasm could be distinguished from the other cytoplasms by a very slight displacement of one band in HindIII digests. The substantial variation in mtDNAs as compared with the apparently minor variation among ctDNA, represents circumstantial evidence that the mtDNA carries the cms trait. In wheat, results similar to those obtained in maize have recently been reported (2). The mt and ctDNAs from normal (fertile) and cytoplasmic male-sterile wheat were studied by EcoRI restriction endonuclease fragment analysis. While the ctDNA patterns disclosed no differences, distinctive patterns were observed between the mtDNA from normal and male-sterile cytoplasms. Similar studies in sorghum have re- vealed a more complex situation (Pring, Conde and Warmke, un- published data). Both mt and ctDNAs from the male-sterile were easily differentiated from their fertile counterparts, and there- fore, these results were of no assistance in assigning the cms trait.

Returning to our studies of maize mtDNAs from male-sterile cyto- plasms, it is instructive to consider our collective results. In the past, identification of sterile cytoplasms has been primarily based on fertility ratings in different inbred backgrounds (20,32). To the present, our laboratories have studied, by restriction enzyme fragment analysis, the mtDNA from more than 25 steriles whose identity had been established by their fertility ratings in various inbred backgrounds. Without exception, the two methods always identified the same cms type. The important lesson is that each cms type is consistently associated with an unique mtDNA. It is not unrealistic to propose that the mtDNA diversity present among the cms types is partially due to the factors responsible for the sterility trait.

Other evidence implicates the mitochondrial genome with the factors responsible for the cms trait. Cytological studies of plants containing the Texas cytoplasm have detected mitochondrial degeneration in the tapetum and middle layer of anthers at the tetrad stage of microsporogenesis while no changes in plastids were observed until late in anther development (35). In contrast, no mitochondrial or plastid alterations were seen in anthers from normal cytoplasm-containing plants. Recently, protein differences from sub-mitochondrial particles and from a partially purified ATPase complex were reported when plants with normal and T cytoplasm were contrasted (36). Finally, considerable evidence now suggests that mitochondria are the target site for the extrachromosomally inherited susceptibility to Southern corn leaf blight which is associated with maize containing the Texas cytoplasm (37-39). Collectively, these studies strengthen the contention that the cms trait is coded by the mitochondrial genome.

Breeding for Extrachromosomally Inherited Traits

Traditional plant breeding techniques have been far more effective in altering nuclear genomes than organelle genomes (40). Two characteristics of organelle genomes are undoubtedly responsible for this distinction. With a few exceptions, cytogenes in higher plants manifest strict maternal inheritance and seem to exhibit little or no genetic recombination (41). These two characteristics may not be mutually exclusive. The application of current plant breeding techniques to cytogenes is best demonstrated by considering the cytoplasmic male-sterility trait. This extrachromosomally inherited trait has been widely exploited in higher plants for producing large quantities of hybrid seed (40). Cms eliminates the often costly procedure of hand emasculating the female parent and, consequently, may substantially reduce the cost of hybrid seed production. In maize and most other plant species, the male-sterile cytoplasm must be introduced by a lengthy backcrossing procedure. In essence, this procedure transfers the desirable nuclear constitution through the pollen parent, to the individual containing the male-sterile cytoplasm. Unfortunately, a considerable amount of time is consumed by multiple generations of backcrossing, and the nuclear constitution of the recurrent parent may be somewhat altered at the completion of the backcrossing program. A more detailed treatment of the use of cytoplasmic male sterility in maize seed production is provided by Duvick (33).

The practical aspects of little or no genetic recombination
in organelle genomes is perhaps best illustrated by a situation
which occurred with maize production in the United States. Prior
to the 1970s, a single male-sterile cytoplasm, Texas or T, was
extensively employed in the production of hybrid maize seed (42).
More than 85% of the maize grown in the United States contained
the T cytoplasm. A disease outbreak of epidemic proportions led
to the discovery that maize with T cytoplasm was highly susceptible
to two leaf diseases, Southern corn leaf blight (Bipolaris maydis,
race T, formerly known as Helminthosporium maydis) and yellow leaf
blight (Phyllosticta maydis) and it was necessary to curtail the
use of T cytoplasm in seed production. So far, disease susceptibility and male sterility have displayed an absolute association.
However, it is not clear if this association is due to close linkage
or a pleiotropic effect. In any event, it was not possible, by
conventional techniques, to recombine the T cytoplasm with other
resistant cytoplasms in an effort to eliminate the deleterious
trait, disease susceptibility. Although other sterile cytoplasms
have been identified and are now being used commercially, the
Texas cytoplasm, less the disease problem, would very likely still
be favored because of its stability, sureness in rendering plants
male-sterile, and its well established and understood fertility
restoration system.

The Mitochondrial Genome and Genetic Engineering

The in vitro transfer of mitochondria from one individual
to another is an aspect of genetic engineering which needs attention. The benefits of this technique can be illustrated by considering an example of practical application. Substantial evidence
now indicates that the cms trait is coded by the mitochondrial
genome. This finding suggests that the transfer of mitochondria
from a cytoplasmic male-sterile donor to a normal (fertile) recipient would introduce the sterility-causing cytogenes into the recipient. However, the simple introduction of a few mitochondria
carrying the cms trait would probably not be sufficient to alter
phenotypic expression because the introduced mitochondria would
very likely be swamped by the recipient's own fertile-type mitochondria. The solution probably lies in the development of a
selection system which would favor those mitochondria carrying the
cms trait. The in vitro transfer of mitochondria would have substantial advantages over conventional plant breeding techniques.
For example, the transfer technique should not alter the nuclear
constitution and should be faster because 5 or 6 backcrossing
generations are eliminated. Current developments in protoplast
technology hold promise for making mitochondrial transfer a
reality in the near future (43).

The Texas cytoplasm situation underscores the need for developing new systems for reconstructing the mitochrondrial genome. Nature has apparently placed rather severe limitations to recombination of these genomes in higher plants. However, the genome size and their likeness to prokaryotic genomes suggest that they could be manipulated by the new recombinant DNA technology practiced in microbial systems. Although the recombining of mtDNAs outside of the cell is feasible with today's technology, two serious problems remain. First, restructured genomes need to be selected for the desired genetic content; this will require the development of selection systems for sorting out the appropriate types. Second, transformation systems are needed for returning restructured genomes to plants in a functional form. These are formidable tasks, but if they can be overcome, a new approach for improving the mitochondrial genome will be available.

Although the small size of the mitochondrial genome makes it a primary candidate for initial efforts in genetic engineering, it also limits its potential. Indeed, the vast majority of traits in higher plants are under the control of nuclear genes. It has been suggested that mtDNA might serve as a vehicle for introducing DNA sequences into plant cells. It is true that mtDNAs do have the capability of inhabiting and functioning in plant cells. However, their informational content and location in the cell is both limited and specific. Nonetheless, it may be worthwhile investigating whether genes normally found in the nucleus could function in a stable manner when associated with the mitochondrial genome. Finally, it is possible that gene sequences associated with the mtDNA can be integrated into the nuclear genome. For the moment, this must remain interesting speculation because of our primitive understanding of the phenomenon involved.

THE S CYTOPLASM OF MAIZE

Several features of the S male-sterile cytoplasm of maize make it an especially attractive model system for exploring the genetic engineering problem. Studies of the S cytoplasm have indicated the existence of plant episomes, and more directly, the occurrence of stable chromosomal integrational events. The discovery of a plant episomal system suggests another approach for the genetic modification of higher plants. In the following sections, we will describe genetic and biochemical studies of the S cytoplasm which have relevance to genetic engineering.

Plasmid-Like DNAs Associated With S Cytoplasm

Biochemical studies of mitochondrial preparations from maize plants containing the S cytoplasm have identified two unique plasmid-like DNAs (44). These plasmid-like DNAs are in addition

to the usual high molecular weight mtDNAs. The two plasmid-like
DNAs have molecular weights of 3.45 (S-F) and 4.10 (S-S) x 10^6 and
apparently exist as linear molecules. Although the linear configuration was established by electron microscopy, it is possible that
the linear molecules were derived from the breakage of native
circular molecules (Figure 3).

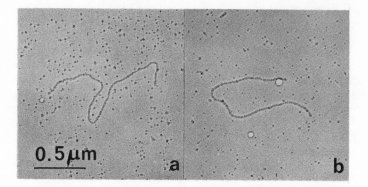

Figure 3. Electron micrographs of linear molecules of the plasmid-
like DNA associated with the S cytoplasm of maize; a) S-S, b) S-F.

The plasmid-like DNAs have been verified in every S cytoplasm
studied regardless of source or nuclear background (44). Thus far,
nine different sources of the S cytoplasm have been examined.
Conversely, the plasmid-like DNAs have not been detected in mtDNA
preparations from normal (fertile) or any of the other male-sterile
cytoplasms. This finding has been repeatedly confirmed among many
sources of these cytoplasms. In addition, the unique DNA species
were not observed in mtDNA preparations from two close relatives of
maize, teosinte and tripsacum. This association between the plasmid-
like DNAs and the S type of male sterility is suggestive of a
causal relationship.
    The plasmid-like DNAs linked with the S cytoplasm have been
successfully isolated exclusively from mitochondrial preparations.
Numerous attempts to obtain these DNAs from chloroplast and nuclear
preparations have failed. Strict maternal transmission of mtDNAs
has been confirmed by restriction endonuclease fragment analyses
(3,18). Similarly, maternal transmission has been established for
the plasmid-like DNAs associated with the S cytoplasm (unpublished
data).
    Electron microscopy investigations of the plasmid-like DNAs
(S-S and S-F) have uncovered an unusual sequence arrangement.
Both the S-S and S-F molecules contain terminal inverted repeats
which are 196 and 168 nucleotides long, respectively (45). The

relevance of terminal inverted repeats to these molecules is not clear. In lower organisms, inverted repeats are frequently involved with insertional events; a similar explanation may apply to these molecules.

Stability of the S Cytoplasm - Cytoplasmic Changes

Laughnan and his associates began several years ago to investigate the stability of the S cytoplasm (46-49). Principally, they looked for cases in which the S cytoplasm changed from the male-sterile to the male-fertile condition. Their endeavors were fruitful in that they turned up several hundred mutations in which male-steriles reverted to fertiles. Significantly, their experimental conditions were appropriately designed for distinguishing the kinds of changes which occurred. Two types of changes were encountered; cytoplasmic mutations from the male-sterile to the male-fertile condition, and nuclear mutations giving rise to new fertility restoring genes. Complete details of the breeding procedures utilized in these investigations are available in Laughnan's papers (46-49).

The bulk of Laughnan's reversions from the male-sterile to male-fertile condition were due to cytoplasmic changes. Well over 300 independently arising cases of cytoplasmic reversions have been established by his test procedures. These revertants originated either as fertile chimeras or completely fertile tassels. Once established, these cytoplasmic revertants have persisted through subsequent generations of propagation. Although these revertants have arisen among several different inbred lines, the majority occurred in a single inbred line which seems especially prone to the event. These results suggest that the nuclear constitution may influence the frequency of the cytoplasmic change.

In collaboration with Laughnan's group, we have recently studied the mtDNA constitution of four newly arisen, male-fertile revertants. In these four stocks, cytoplasmic changes were responsible for the reversion to the male-fertile condition. For comparison, we also studied nonrevertant, male-sterile members of the same families in which the reversion to male fertility occurred. The mtDNAs were isolated from various types and fractionated by gel electrophoresis. Normally, the S cytoplasm contains the two plasmid-like DNAs in stoichiometric amounts. Laughnan's nonrevertant and revertant stocks both deviated from this expectation.

The plasmid-like DNAs, S-S and S-F, were no longer present in those stocks which had reverted to the male-fertile condition by cytoplasmic change (Laughnan, Gabay, Levings, Pring and Conde, unpublished data). This was true for each of the four revertants investigated. Seemingly, the loss of the S-S and S-F DNAs was correlated with the change from the male-sterile to male-fertile phenotype. This result constitutes additional strong evidence that

the S type of male sterility is associated with S-S and S-F DNAs.
The relationship between the S-S and S-F DNAs and mtDNA is not
clear, although it is tempting to consider their relationship as
analogous with that of the bacterial chromosome and its plasmids.
The fate of S-S and S-F DNAs in the revertants is not determined by
these studies. Two possiblities are being considered: the plasmid-
like DNAs may have been lost from the organism or they may have
been transposed to another site.

When we examined the mtDNA constitution of the male-sterile
(nonrevertant) members of the same families in which the reversion
to male fertility occurred, a surprising change was discovered.
Although S-S and S-F DNAs were both present, the S-F DNA was
present in substantially reduced amounts as compared with normal
S cytoplasm. It appeared that although the quantity of S-F DNA
was diminished, the amount of S-S DNA remained unchanged. These
findings suggest that the disappearance of the S-F DNA is a prelude
to the complete loss of both the S-S and S-F DNAs and the concomitant
reversion from male sterility to male fertility. The instability
of these populations apparently stems from the elimination of the
S-F DNA species by an unknown mechanism. The reason for the subsequent loss of the S-S DNA species is not clear, but it is not
unreasonable to speculate that S-S DNA replication is in some way
dependent on the presence of the S-F molecule. Finally, the purg-
ing of the S-S and the S-F DNA from their association with mtDNA
seems to be influenced by the nuclear genome. This is indicated
by the fact that certain nuclear backgrounds have much higher frequencies of reversion from the male-sterile to the male-fertile
condition.

Stability of the S Cytoplasm - Nuclear Changes

A second class of male-fertile reversions was identified by
Laughnan's group. Test cross analyses indicated that the change
did not happen at the cytoplasmic level, but instead, the new male
fertiles exhibited a behavior expected of nuclear restorer genes.
So far only 10 new restorer strains have been identified and, to
some extent, characterized. The new restorers first occurred as
either fertile chimeras or completely fertile tassels.

The naturally occurring restorer of cms-S is the Rf3 locus.
The 10 new restorers have a gametophytic mode of restoration which
is also true of the standard S restorer, Rf3. However, the 10
differed from the standard Rf3 locus in several aspects. For
example, the new restorers often exhibited reduced transmission
through the female gametophyte, a reduction in kernel size, and a
lethality of the restorer homozygote. Especially interesting was
one of the restorers, designated as IV, that, unlike the others,
originated in a maintainer plant and seemed free of adverse effects.
Although these distinctions hinted that the new restorers were not

simply mutations of the standard Rf3 locus, it remained for mapping
studies to provide the conclusive evidence.

The standard restorer locus, Rf3, has been mapped in the long
arm of chromosome 2. Roman numeral designations have been assigned
to the 10 new restorer genes. Mapping studies have placed restorers
I and VIII on chromosome 8, IV and VII on chromosome 3 and IX and
X on chromosome 1. Even though the remaining four restorers, II,
III, V, and VI, have not been unequivocally mapped, they have been
verified as nonallelic with Rf3. Very surprisingly, the 10 new
restorers apparently each occupied unique chromosomal locations.

These results prompted the investigators to propose a male
fertility element with episomal characteristics. Within bacterial
systems, episomes are capable of being transposed from one site to
another or of being entirely lost. The transposing phenomenon is
suggested by the origin of the newly arisen restorer genes. Seemingly, the new restorer arose by the integration of fertility-
restoring elements at unique chromosomal sites. As indicated earlier,
the new restorer often exhibited erratic behavior. The investigators have speculated that this aberrant behavior may result from
either differences in integration sites on the chromosomes or in
qualitative differences in the fertility elements. The latter
explanation is advocated by the fact that restorer IV, the only one
of the restorers that has not displayed deleterious side effects, is
the only one which originated in a fertile maintainer cytoplasm.

The investigators have reasoned that the male-fertile revertants
have a common origin even though the changes occurred at both the
nuclear and cytoplasmic levels. This contention is supported by
the fact that the two kinds of male-fertile exceptions have originated in the same strains and in both instances are expressed first
as either complete male-fertile tassels or as fertile-sterile tassel
chimeras. Finally, they propose that the male fertility element
is fixed in the cytoplasm when a cytoplasmic change from the male-
sterile to the male-fertile condition transpires. On the other hand,
if the element is fixed in the nucleus, it behaves as a restorer strain.

Genetic Engineering With Plant Plasmids

Although the investigations of the S cytoplasm are still incomplete, the picture which seems to be developing is that of a
plasmid-like entity in a higher plant system. The relationship
between the mtDNA and the unique DNA species, S-S and S-F, appears
remarkably similar to the arrangement between bacterial chromosomes
and plasmids. This comparison is not necessarily intended to imply
common origin, but rather to suggest that they may have similar
strategies. The genetic and biochemical data indicate that the
plasmid-like DNAs coexist with the mitochondrial genome, and more
importantly, exercise control of a phenotypic trait. Of extraordin-
ary significance is the fact that the evidence also suggests these

plasmid-like DNAs are capable of transposition. The indication
that these DNAs may be integrated into the nuclear genome and
effect a phenotypic alteration is especially relevant to genetic
engineering. The prospect that this system could be manipulated
for the modification of phenotypic expression in higher plants is
very exciting. Obviously, the feasibility of this approach is
still obscure, but several positive aspects are noteworthy.
Recent advances in recombinant DNA research, particularly in
bacterial systems, and improvements in protoplast and tissue
culture techniques, are providing some of the very methodologies
required for this approach. These successes predict that genetic
engineering through plant plasmids will soon be examined.

Acknowledgment: Part of the research reported here was supported by grants from the National Science Foundation (PCM 76-
09956) and Pioneer Hi-Bred International, Inc.

REFERENCES

1   Borst, P. (1977) in International Cell Biology 1976-1977
    (Brinkley, B.R. and Porter, K.R., eds.), pp. 237-244,
    Rockefeller Univ. Press, New York, NY.
2   Quetier, F. and Vedel, F. (1977) Nature 268, 356-368.
3   Pring, D.R. and Levings, C.S. III. (1978) Genetics 89, 121-136.
4   Pring, D.R. and Conde, M.F. (1978) Plant Physiol. 61 (Suppl),
    114.
5   Sisson, V.A., Brim, C.A. and Levings, C.S. III. (1978)(in press).
6   Kolodner, R. and Tewari, K.K. (1972) Proc. Nat. Acad. Sci.
    U.S.A. 69, 1830-1834.
7   Vedel, F. and Quetier, F. (1974) Biochem. Biophys. Acta 340,
    374-387.
8   Shah, D.M., Levings, C.S. III, Hu, W.W.L. and Timothy, D.H.
    (1976) Maize Genet. Coop. Newsletter #50, 94-95.
9   Shah, D.M., Levings, C.S. III, Hu, W.W.L. and Timothy, D.H.
    (1979) (submitted).
10 Synenki, R.M., Levings, C.S. III and Shah, D.M. (1978) Plant
    Physiol. 61, 460-464.
11 Hayward, G.S., Jacob, R.J., Wadsworth, S.C. and Roizman, B.
    (1975) Proc. Nat. Acad. Sci. U.S.A. 72, 4243-4247.
12 Bernard, U., Bade, E. and Küntzel, H. (1975) Biochem. Biophys.
    Res. Commun. 64, 783-789.
13 Terpstra, P., Holtrop, M. and Kroon A.M. (1977) Biochem. Biophys.
    Acta 475, 571-588.
14 Morimoto, R., Lewin, A., Hsu, H.J., Rabinowitz, M. and Fukuhara,
    H. (1975) Proc. Nat. Acad. Sci. U.S.A. 72, 3868-3872.
15 Sanders, J.P.M., Heyting, C., Verbeet, M.P., Meijlink, F.C.P.W.
    and Borst, P. (1977) Mol. Gen. Genet. 157, 239-261.
16 Prunell, A., Kopecka, H., Strauss, F. and Bernardi, G. (1977)
    J. Mol. Biol. 110, 17-52.

17   Stepien, P.P., Bernard, U., Cooke, H.J. and Küntzel, H. (1978)
      Nucl. Acids Res. 5, 317-330.
18   Levings, C.S. III and Pring, D.R. (1976) Science 193, 158-160.
19   Levings, C.S. III and Pring, D.R. (1977) J. Hered. 68, 350-354.
20   Sanders, J.P.M., Borst, P. and Weijers, P.J. (1975) Mol. Gen.
      Genet. 143, 53-64.
21   Gracen, V.E. and Grogan, C.O. (1974) Agron. J. 65, 654-657.
22   Levings, C.S., III and Pring, D.R. (1976) Proc. 31st Ann. Corn
      Sorghum Res. Conf. pp. 110-117.
23   Conde, M.F., Pring, D.R. and Levings, C.S. III (1978) (submitted).
24   Levings, C.S. III, Timothy, D.H., Pring, D.R., Conde, M.F. and
      Kermicle, J.L. (1979) (submitted).
25   Leaver, C.J. and Harmey, M.A. (1973) Biochem. Soc. Symp. 38,
      175-193.
26   Pring, D.R. (1974) Plant Physiol. 53, 677-683.
27   Leaver, C.J. and Harmey, M.A. (1976) Biochem. J. 157, 275-277.
28   Cunningham, R.S. and Gray, M.W. (1977) Biochem. Biophys. Acta
      475, 476-491.
29   Leaver, C.J. (1976) in Genetics and Biogenesis of Chloroplasts
      and Mitochondria (Bücher, T., Neupert, W., Sebald, W. and
      Werner, S., eds.), pp. 779-782, North-Holland, New York, NY.
30   Prunell, A. and Bernardi, G. (1974) J. Mol. Biol. 86, 825-841.
31   Edwardson, J.R. (1970) Bot. Rev. 36, 341-420.
32   Beckett, J.B. (1971) Crop Sci. 11, 724-727.
33   Duvick, D.N. (1965) Adv. Genet. 13, 1-56.
34   Levings, C.S. III and Pring, D.R. (1979) in Physiological
      Genetics (Scandalios, J.G., ed.), Academic Press, New York,
      NY (in press).
35   Warmke, H.E. and Lee, S.L.J. (1977) J. Hered. 68, 213-222.
36   Barratt, D.H.P. and Peterson, P.A. (1977) Maydica 22, 1-8.
37   Miller, R. and Koeppe, D.E. (1971) Science 173, 67-69.
38   Peterson, P.A., Flavell, R.B. and Barratt, D.H.P. (1975)
      Theor. Appl. Genet. 45, 309-314.
39   Aldrich, H.C., Gracen, V.E., York, D., Earle, E.D. and Yoder,
      O.C. (1977) Tissue Cell 9, 167-178.
40   Harvey, P.H., Levings, C.S. III and Wernsman, E.R. (1972)
      Adv. Agron. 24, 1-27.
41   Sager, R. (1972) Cytoplasmic Genes and Organelles, p. 405,
      Academic Press, New York, NY.
42   Ullstrup, A.J. (1972) Annu. Rev. Phytopathology 10, 37-50.
43   Carlson, P.S. (1973) Proc. Nat. Acad. Sci. U.S.A. 70, 598-602.
44   Pring, D.R., Levings, C.S. III, Hu, W.W.L. and Timothy, D.H.
      (1977) Proc. Nat. Acad. Sci. U.S.A. 74, 2904-2908.
45   Levings, C.S. III, Hu, W.W.L., Timothy, D.H. and Pring, D.R.
      (1979) (submitted).
46   Laughnan, J.R. and Gabay, S.J. (1973) Theor. Appl. Genet. 43,
      109-116.
47   Laughnan, J.R. and Gabay, S.J. (1975) in Genetics and Biogenesis
      of Mitochondria and Chloroplasts (Birky, C.W., Perlman, P.S. and
      Byers, T.W., eds.), pp. 330-349, Ohio State Univ. Press, Columbus.

48   Laughnan, J.R. and Gabay, S.J. (1975) in International Maize
      Symposium: Genetics and Breeding (Walden, D.B., ed.), John
      Wiley and Sons, Inc., New York, NY (in press).
49   Singh, A. and Laughnan, J.R. (1972) Genetics 71, 607–620.

HOST-VECTOR SYSTEMS FOR GENETIC ENGINEERING OF HIGHER PLANT CELLS

C. I. Kado

Department of Plant Pathology
University of California
Davis, California 95616

INTRODUCTION

Methods for the insertion of exogenous genetic material into higher plants have been long sought, with the hope that some of those inserted genes will be able to cause genetic transformation. Earlier reports of successful incorporation of exogenously applied DNA and bacteriophages to intact plants were later shown to be the result of artifacts or the experiments remained irreproducible (1-4). Nevertheless, such reports have stimulated other workers to investigate the possibility of transforming plant cells. If plants could be manipulated and modified genetically by laboratory methods, then new genetic variations might be generated much faster than by classical sexual processes. More importantly, this technique might permit the broadening of the genetic pool.

There are a number of limitations in the above premise. First, it must be realized that genetic elements foreign to the higher cell are subject to scrutiny by any one of several higher cell surveillance systems, one of which is the genetic barrier. Other inherent barriers are mechanical, biochemical, and biophysical in nature. Second, the genetic elements must be stably maintained, replicated in the plant, and able to be passed on stably through seeds. Thirdly, such elements must be expressed into desirable phenotypic features. Fourthly, the desirable genes might contribute to unpredictable, undesirable features that would lead to crop vulnerability, such as susceptibility to new plant diseases, pests, and environmental factors. Finally, if such plants were developed, it is uncertain whether cost factors would be prohibitive for commercial production.

In this chapter, current vectors and techniques to circumvent some of these mechanical and biochemical barriers are described.

223

Several vector molecules that may prove potentially useful in the insertion of foreign genetic material, are also described. Each has its own inherent limitations and these will be pointed out.

For the insertion technique, the classical concepts of transformation and transfection are followed but differ in that these concepts, commonly aligned with bacteria, are transposed to higher plant cells.

BARRIERS TO FOREIGN GENES

It is of current concern to determine whether or not foreign genetic elements can cross the genetic barriers, be expressed, and regulate the host cell so that phenotypic features can be measured. The idea that cells possess mechanisms that enable them to reject foreign nucleic acids and thus help them survive with an unchanged genetic structure is significant.

Introduction of bacterial DNA by incubating seedlings, seeds and callus tissues with purified preparations of bacterial DNA has been attempted (1-4). In all these trials, the exogenously-added DNA was degraded extensively, indicating that various plant nucleases serve as a protective biochemical barrier. The nuclease barrier is, therefore, one of the first encountered in DNA uptake studies. Exogenously-added DNA is also bound to cell wall components, often in a pancreatic deoxyribonuclease-resistant form. A DNA binding protein, particularly rich in seed cotyledons, will maintain the DNA in a stable form for a considerable time (Kado, unpublished data). This mechanical barrier can contribute to erroneous interpretations of DNA uptake unless precautionary measures are taken -- there may be nonspecific adsorption of the exogenously-added DNA to cell wall material, which immobilizes the DNA, and is either released during isolation of DNA from the treated cells or measured as uptake of radiolabeled DNA by whole cells.

If the exogenous DNA finds its way into the cytoplasm in a relatively intact state, it may still be bombarded by several enzymes. To date, no restriction endonucleases have been found in higher plants but this does not mean that other DNA modifying enzymes are not present. Indeed, plasmid DNA in the cytoplasmic fraction undergoes considerable degradation (5). Nevertheless, if a certain number of intact molecules persist and make their way to the nuclei, the nuclear membrane itself may prove to be a second mechanical barrier, even though nucleopores are numerously scattered about and might serve as portals of entry for the foreign genetic material. DNA may be prevented from gaining entry into the nucleoplasm by adsorption to the nuclear membrane. However, this barrier does not seem to be as formidable as the cell wall and its membranes. It has been shown that single-stranded DNA is adsorbed more tenaciously into the nuclear membrane than is double-stranded DNA (6). Finally, if the exogenous DNA finds its way into the nucleoplasm, it might encounter more formidable genetic barriers. DNA sequence recognition sites may be foreign to the plant cell, permitting no processing of the introduced genetic material. Integration of foreign genes into some chromosomal element is a prerequisite for their survival through meiosis. Integration may not be necessary if the genetic element can carry with it its own replication machinery and be an autonomous replicon, like its plasmid counterpart in bacteria. Plants generated from these cells can be propagated vegetatively so that perpetuation of genetically engineered plants through seed is not mandatory.

POTENTIAL VECTORS

Potential vectors must have specific attributes before they can be considered for use as cloning vehicles. The vector must contain a replicon capable of being recognized and stably maintained indefinitely in a primary vector-generating host cell, as in the case of plasmid and bacteriophages in bacteria. It must be replicated into multiple copies (amplified) to yield sufficient quantities of genetic material to handle economically. It must be easy to purify without degradation. It must also be able to be inserted, maintained and replicated in the plant cell without appreciable degradation.

Several vectors fit some but not all of these criteria. Continued efforts in search of vectors fitting all of these prerequisites are needed.

VIRUSES

Plant viruses, particularly those with DNA genomes, are one choice of vectors made by various laboratories. Several DNA plant viruses have been characterized (Table 1), of which cauliflower mosaic virus is the best known for its cloning vehicle potential. Purified preparations of this viral DNA can infect crucifers with relatively good efficiency (7). It can be assumed that DNA from the other viruses listed in Table 1 will also prove to be infectious. These virions, therefore, can serve as possible transducing vehicles for the introduction of foreign DNA segments into plant cells.

The limited host-range of cauliflower mosaic virus makes it less attractive than the related caulimovirus, Dahlia mosaic virus, which infects a number of species in four plant families (8), several of which are important crop plants.

Because the yield of these virions in plants is extremely low, leading to even lower yields of their DNA, segments of the cauliflower mosaic virus DNA have been cloned in strains of restriction/ modification/recombination-deficient E. coli. Restriction endonucleases, SalI and BamH1, cleave cauliflower mosaic virus DNA at a single site, generating a linear molecule about 7500 base pairs long (9). Therefore, virtually the entire viral genome can be cloned. This has been done with SalI-generated molecules ligated

Table 1

Viruses as Vectors of DNA Sequences

Virus	Size of virion (μm)	Size of genome (mol. wt.)	Plant host-range	Ref.
Cauliflower mosaic virus	50	4.4×10^6	Cruciferae	7
Carnation etched ring virus	42	?	Datura stramonium Dianthus caryophyllus Silene armeria	13,14 17
Cassava latent virus	15–20	0.8×10^6	Cassava	20
Dahlia mosaic virus	45	?	Amaranthaceae Chemopodiaceae Compositae	18 8
Golden yellow mosaic virus	18	$0.66\text{–}0.95 \times 10^6$	Solanaceae Leguminosae	19
Maize streak virus	15–20	0.71×10^6	Maize	20
Mirabilis mosaic virus	45–50	?	Nyctaginaceae (Mirabilis sp.)	21
Potato leafroll virus	25	0.56×10^6	Amaranthaceae Solanaceae	22 24
Strawberry vein banding virus	45–50	?	Rosaceae (Fragaria sp.)	23
λgt	54(head);150(tail)	25.9×10^6	E. coli	25,27,28
M13	800 x 6	2×10^6	E. coli	29

to plasmid pGM706. Infectivity tests of the recombinant molecule in tender green mustard, after propagation in E. coli and religation, proved to be negative (9). In fact, mere exposure of the viral DNA to SalI, followed by religation, caused loss of infectivity. The reasons for this remain unknown. Thus, based on these results, it is not certain whether or not cauliflower mosaic virus will be useful as a vector.

Insertion of new genetic material into the viral genome by enzymatic gene-splicing techniques can be accomplished. However, besides infectivity inactivation by restriction nuclease, one or more barriers can be foreseen. The host-range of cauliflower mosaic virus is restricted to crucifers. Dahlia mosaic virus is probably better suited than cauliflower mosaic virus because its host-range extends beyond the Cruciferae. Caulimoviruses are likely to be replicated in the cytoplasm of the host cell rather than in the nucleus (10-12), although Favali et al. (15) have proposed the nuclei as the primary site of the replication of cauliflower mosaic virus based on differences in silver grain distributions between infected and uninfected cells. However, virus perturbed cells are known to possess enhanced nucleic acid synthesis. Therefore, possible integration of foreign genes with caulimovirus DNA into host nuclear material seems remote.

In vitro splicing of genes on segments of caulimoviral genomes that carry its origin of replication may prove to be a means of circumventing these restrictive barriers. Various recombinants of cauliflower mosaic virus DNA have been cloned using these different plasmid vectors: ColEl, pMB9 and pBR313 (16). However, it is still unknown if caulimoviral DNA segments in recombinant molecules will replicate and integrate in the nucleoplasmic material.

The use of caulimoviral recombinant molecules for direct inoculation in a suitable host does not provide a means for selection. It will be extremely difficult to delineate between infection by recombinant molecules and nonrecombinant molecules. Besides, it is uncertain at the present time whether the complete caulimoviral genome is mandatory for infection. It has already been stated (9) that single-site cleavage with SalI and rejoining of cauliflower mosaic virus DNA does not restore infectivity.

Nascent infections (DNA replication without symptoms usually scored by visual inspection) by caulimoviral DNA recombinants may occur upon direct inoculation by in vitro preparations of these recombinants. It is nearly impossible to discriminate these infected cells from adjacent, uninfected cells unless the host has some distinguishing features which are expressed when infection takes place.

A means of avoiding this problem is to work with single plant cells. This permits direct manipulation of host cells on a one-to-one basis and the fate of the caulimoviral recombinant DNA molecule can then be followed. Methods facilitating such studies have been described (5,59).

Other plant DNA viruses, such as potato leafroll virus, have the inherent disadvantage of producing very low yields of virus from

infected sources. At the present time, no work has been done on
restriction endonuclease cleavage patterns of its DNA so that single
restriction sites are unknown.

Besides plant DNA viruses, alternative virus vector systems
might be considered. For example, bacteriophage lambda mutants,
known as λgt-λC, have been constructed by Thomas et al. (28). These
mutants carry two EcoRI restriction sites between which EcoRI-
generated DNA segments of 1 to 14 kilobase pairs can be inserted.
The λgt-λC mutants themselves are unable to produce plaques in the
E. coli cloning host. However, plaques are produced if a new DNA
segment within the above DNA size range is inserted between the two
EcoRI sites because sufficient length of DNA is apparently necessary
for packaging into viable phage particles (28). Thus observation
of plaques after insertion of new DNA constitutes a powerful positive
selection system. Davis et al. (30) have recently shown that λgt
mutants containing a particular size of DNA can be selected either
physically, by their buoyant density (26), or genetically, by growth
on pel E. coli (31,32), a host that is very stringent for phage-
containing DNA of wild-type size or larger. These lambda mutants
possessing varying amounts of DNA in the vector portion can therefore be used to select specific sizes of DNA from a random mixture
of DNA to be cloned. The efficiency of this selection is extremely
high (up to 1). Methods for screening these clones have been
developed using plaque hybridization (33,34), in situ hybridization
(35), and immunological procedures (36). Davis et al. (30) have
recently developed a simple technique to screen large numbers of
clones. Plaques usually contain free phage DNA. A dry nitrocellulose filter, placed in contact with the plaques, permits rapid
adsorption of the free DNA. Up to 2×10^4 plaques in a single petri
plate can thus be screened. After the filter has been in contact
with the plaques for about 5 min, the DNA is denatured with
alkali, and the filter washed and dried. It is then incubated with
radioactive complementary nucleic acid, washed, dried and analyzed
by autoradiography on x-ray film. Phage harboring the desired
cloned DNA can then be spotted on the autoradiogram and selected
from the corresponding plaque. Virtually any gene within the size
limitations of insertion in λgt can be cloned. The EcoRI sites
themselves can be altered to suit the fragment being cloned by
attaching adapter fragments to the cloning vector. Procedures for
use of adapter or linker molecules have been described (37-39).

Recently, the filamentous coliphage M13 has been introduced as
a potential cloning vehicle (40). Because wild-type M13 is a single-
stranded circular DNA, the double-stranded supercoiled replicative
form RFI is used for cloning. A nonessential region of this phage
has been located and a hybrid phage containing the HindII fragment
from the lac regulatory region of the lac operon and the part of the
β-galactosidase gene that codes for the α-peptide has been constructed (40). This hybrid phage (M13mp1) is plated with E. coli
K12 71-18 (Δ[lac, pro], F' lac IqZΔ M15 pro+) in soft agar containing isopropylthiogalactoside (IPTG), an inducer of the lactose

operon, and 5-bromo-4-chloro-indolyl-β-D-galactoside, a colorless
compound which, when hydrolyzed by β-galactosidase, releases deep
blue 5-bromo-4-chloroindigo (41). Thus, infection of E. coli by
M13mp1 is reflected by the formation of blue plaques. On the other
hand, in vitro insertion of foreign DNA in the lac region of this
phage results in the loss of α-complementation and the appearance
of white plaques. These plaques can be readily distinguished from
the blue ones, which contain rejoined parental phage molecules.

PLASMIDS

     Theoretically any plasmid may serve as a vector for foreign
DNA but because of possible hypothetical dangers which may arise
through the use of promiscuous conjugative plasmids as cloning
vehicles, only a few nontransmissible plasmids have been employed
in current recombinant DNA methodologies in the United States.
The list of plasmids in Table 2 may serve as examples. Each of
these plasmids has certain features which are desirable for recombinant DNA work. They carry genes for easy identification (antibiotic resistance), high production (amplification under relaxed
control), uncomplicated insertion (single restriction sites), and
are usually of minimal size to avoid background noise. Plasmids
pBR313 and pBR322 are illustrated as some of the more efficient
vectors because both plasmids have genes for conferring ampicillin
and tetracycline resistance. Bolivar et al. (49,50) pointed out
the advantages of using these reconstructed plasmids as cloning
vehicles. Single restriction endonuclease sites for EcoRI, HindIII,
BamHI, SalI, HpaI and SmaI in pBR313 are available and an additional
PstI site in the ampicillin gene has been constructed in pBR322.
This permits the cloning of foreign DNA with each of these restriction enzymes so as to increase the probability of inserting a fully
functional piece of foreign DNA (e.g., one enzyme may cleave in
the desired cloned gene whereas the cleavages of another enzyme
would flank that gene). The substrate sequences for HindIII, BamHI
and SalI restriction enzymes are in the gene conferring tetracycline
resistance of E. coli. Since the insertion of foreign DNA using any
one of these restriction enzymes leads to the inactivation of tetracycline resistance (i.e., insertional inactivation), the recovery of
cells harboring recombinant molecules is greatly facilitated by
selecting ampicillin-resistant (Ap^R), tetracycline-sensitive (Tc^s)
phenotypic clones in the following manner. Ap^R, Tc^s transformants
can be enriched by taking advantage of the bacteriostatic nature of
tetracycline and the bactericidal effects of ampicillin and cycloserine. The growth of Tc^s recombinant transformants is inhibited
by the addition of tetracycline to the medium (45 min exposure to
10 μg/ml tetracycline at 37°). Cycloserine is then added (100 μg/ml,
1 hr, 37°) to promote the lysis of any growing cells. Cells containing recombinant DNA can then be recovered by washing away the
antibiotics. Untransformed cells can be eliminated from the culture

Table 2

Plasmid Cloning Vehicles

Plasmid	Molecular weight	Organism	Copies per cell[a]	Antibiotic markers[b]	Single restriction endonuclease site	Ref.
pSC101	5.8×10^6	E. coli	1-2	Tc	EcoRI BamHI SalI	42,43
ColE1	4.2×10^6	E. coli	1000-3000^{Cm}	Col E1-producing, Col E1		44
RSF2124	7.3×10^6	E. coli	280^{Cm}	Ap	EcoRI	45
pVH51	2.1×10^6	E. coli	935^{Cm}	Col E1	EcoRI	46
pGM16	12.1×10^6	E. coli	relaxed^{Cm}	Tc, Km, Col E1	BamHI	43
pCR1	8.7×10^6	E. coli	relaxed^{Cm}	Km	EcoRI	47
pMB9	3.5×10^6	E. coli	relaxed	Ap, Tc	EcoRI, HindIII, BamHI, SalI	48,49
pBR313	5.8×10^6	E. coli	relaxed^{Cm}	Col E1, Ap, Tc	EcoRI, SmaI, HpaI, SalI, HindIII, BamHI	
pBR322	2.6×10^6	E. coli	relaxed^{Cm}	Col E1, Ap, Tc	EcoRI, HindIII, BamHI, SalI, PstI	50
RK2::Mu	$28.9\text{-}42.7 \times 10^6$	Gram negatives	stringent	Ap, Km, Tc	EcoRI	51
PUB110	3×10^6	Bacillus subtilis	1000	Km	EcoRI, XbaI, BamHI, BglII	52
PSC194	4.9×10^6	Staphylococcus aureus	stringent	Sm, Cm	EcoRI	53,54
Scp1	9.0×10^6	Saccharomyces cerevisiae	50-100	Om(?), Vm(?)	PstI, HpaI	94

[a] Relaxed control induced by chloramphenicol^{Cm} treatment results in high copy numbers (amplification).

[b] Abbreviations (other antibiotics): Ap, ampicillin; Col E1, colicin E1; Km, kanamycin; Om, oligomycin; Sm, streptomycin; Tc, tetracycline; Vm, venturicidin. All markers except Col E1-producing are antibiotic resistance markers.

by the addition of ampicillin (20 μg/ml, 4 to 12 hr, 37°) either
before or after these steps. Finally, the recombinant molecule is
amplifiable by the addition of chloramphenicol to the medium (180
μg/ml, 37°, overnight) so that high yields of recombinant molecules
can be obtained. Consequently, cloned DNA segments are available
in quantities amenable to DNA sequencing (e.g., by the method of
Maxam and Gilbert (55)). The plasmid pBR322 itself has already
been sequenced (G. Sutcliffe and W. Gilbert, personal communication;
97) thereby adding to the advantage of using this vector.

Owing to the above advantageous properties of pBR313 and pBR322,
studies on the means of inserting these plasmids have been under-
taken (56-59). Furthermore, the optimal conditions for the uptake
of these plasmid vectors, including ColE1, pCR1 and the Agrobacterium
Ti-plasmid, by plant protoplasts have been worked out (5,56-59).
These vectors can be detected in the nuclei of viable protoplasts
but no longer retain their supercoiled structure (58,59). All of
these plasmids will undergo depolymerization in the cytoplasm but
the degree of depolymerization depends on the source of protoplasts.
Tests made on several plant protoplast sources indicate that cowpea
and especially turnip mesophyll are low in depolymerizing enzymes
and therefore maintain pBR313 molecules in stable form (59). Recombinants of the Agrobacterium Ti-plasmid and pBR313 and pBR322
have been inserted into cowpea protoplasts and preliminary indications are that bonafide transcripts are synthesized in these protoplasts (60). Biological expression of foreign genes in plant
protoplasts needs to be firmly assessed. Difficulties have been
encountered in regenerating low nuclease protoplasts on medium for
positive selection of protoplast clones. Also, only a few useful
markers for positive selection have been developed as protoplast
sources (e.g., the streptomycin- or valine-resistant lines of
tobacco (Nicotiana tabacum)) (61-63). When such markers are well
characterized, these genes can be cloned using one of the available cloning vehicles described above. The use of eukaryotic
plasmid Scp1 of yeast as the cloning vector may prove to be more
functional than the bacterial plasmid vectors. Eukaryotic genes
may be cloned using Scp1 in an eukaryotic background. A transformation system has been recently developed for yeast (95;
this volume).

The problem confronting plant scientists is the lack of any
biochemical mutants of higher plant cells that can be used for
direct complementation assays. Such assays can be performed
indirectly using plant DNA segments cloned in E. coli deficient in
a particular enzyme. Of course, the basic assumption is made that
the eukaryotic enzyme counterpart functions in E. coli. It has
been shown that cloned yeast (S. cereviseae) DNA, inserted into
and E. coli hisB deletion mutant lacking imidazole glycerol phosphate dehydratase activity is detected in the cell extracts when
the yeast sequence is integrated into the mutant chromosome so that
complementation by an eukaryotic sequence actually functions. Complementation of leuB (β-isopropylmalate) dehydrogenase, trpAB

(tryptophan synthase), and <u>argH</u> (argininosuccinate lyase) in <u>E. coli</u> by the yeast genes have also been reported (96). Thus the isolation and cloning of plant genes with functional counterparts in deletion mutants of <u>E. coli</u> or <u>S. cereviseae</u> is a powerful alternative complementation assay. In the end, however, biochemical mutants of plants will be essential for direct proof that cloned plant structural genes are functioning.

BACTERIA

Agrobacterium tumefaciens harbors large and small plasmids (67-69) of which one, averaging 120 megadaltons, confers virulence on the organism. Current evidence suggests that this organism might be inserting part of its genetic material into plant DNA as a plasmid segment during infection of dicotyledonous plants. The potentials of this organism for genetic modification of plants are discussed in another chapter of this volume. The concept of using the Agrobacterium plasmid as a vector for foreign genes has been enhanced recently by the detection of DNA sequences in tobacco tumor calli that are complementary to certain Agrobacterium plasmid segments generated by restriction enzymes (70,71). RNA derived from these calli were shown to hybridize with those specific plasmid segments (72,73). At the present time, it is not known if the putative Agrobacterium plasmid sequences are integrated, although recent data by de Picker et al. (74) are highly suggestive that integration does occur. Indirect evidence, such as long term maintenance of those sequences in tobacco calli, has suggested the possibility of integration with plant chromosomal elements. It is still unclear if only a small segment or the entire plasmid itself is inserted into plant cells, and whether or not the segment ends up in the plant nuclei. It has been shown that the Agrobacterium plasmid can be taken up by cowpea protoplasts by first reconstituting the plasmid DNA with the capsid protein of tobacco mosaic virus (56). The plasmid is detected in a somewhat depolymerized state (6-8S molecules) in the nuclei of these protoplasts. However, RNA obtained from these protoplasts will form specific hybrids with the plasmid DNA (60).

Attempts are being made to demonstrate direct biological transformation of protoplasts with the Agrobacterium plasmid. So far, the reports of such transformation have been cautious and have been based on equivocal experimental results. Two of the phenotypic features employed for scoring transformation are the presence of lysopine dehydrogenase and autotrophy to phytohormones. Both phenotypic characters have relatively weak foundations even though positive correlations between octopine/nopaline utilization in culture media and their biosynthesis in crown gall tissues have provided indirect supporting evidence (75,76). Genes involved in the degradation of octopine or nopaline in Agrobacterium are different from those involved in their synthesis in crown gall tissues

(76,77). Goldmann (78) states that this fact argues against the gene transfer hypothesis. A demonstration that biosynthetic enzymes for octopine/nopaline synthesis are direct gene products of Agro-bacterium is needed. The preliminary report of detecting octopine and octopine dehydrogenases in bean embryonic tissues (79) has cast further doubt on the gene transfer hypothesis.

The phytohormone autotrophic phenotype is not an exclusive phenomenon of crown gall cells. Normal cells can be converted by simple selection on hormone-free media to grow like crown gall cells. Habituated normal cell lines are well known for their phytohormone autotrophy (80). Also, tissues subjected to high auxin levels (81) or to aminofluorenes (82) convert to phytohormone autotrophic types.

The idea of employing Agrobacterium as a vehicle for inserting foreign genes through the infection process must be considered judiciously. Of more immediate promise is the concept of ancillary growth on roots using avirulent Agrobacterium (83), that reside on root surfaces and in the rhizosphere, and release nutrients to crop plants. Such an organism can be modified genetically to carry genes that promote plant growth. If indeed Agrobacterium has the means to insert part of its genetic material into plant cells, then part of the problem of overcoming the genetic barriers discussed earlier may be solved.

TRANSPOSABLE ELEMENTS

Transposable elements are DNA segments that can insert into any number of sites in a genome and usually carry genes besides those DNA sequences needed for insertion. These elements, known as transposons (84), can join segments at DNA sites lacking genetic homology, culminating in illegitimate recombination as opposed to genetic recombination involving homologous DNA sequences. Trans-posons vary tremendously in size and genetic complexity, ranging from 2 to 47 kilobase pairs (λ and Mu are the known larger ones). Upon insertion into a genome, they can influence the activity of genes near their sites of insertion and may cause chromosome aberrations. Transposons in bacteria coding for resistance to antibiotics have been described (85-92) and were found originally as cointegrates of conjugative plasmids (Table 3).

These transposition elements, inserted into desired DNA regions, serve as vectors with assayable markers. Thus they can be intro-duced into plant protoplasts, which can in turn be selected on the basis of resistance to the antibiotic conferred by the transposon. For example, Tn5 confers resistance to kanamycin. Protoplasts are usually sensitive to this antibiotic so that a positive selection system is possible, although expression of antibiotic-resistance markers of transposition elements in plant protoplasts has not been studied in depth. Nevertheless, these elements, when inserted next to genes to be studied (based on known genetic maps), serve as tags

Table 3

Transposable Element Vectors

Transposon	Size (kb)	Originally found in plasmid	Antibiotic resistance[a]	Insertion specificity	Ref.
Tn1	4.8	RP4	Ap	random	85
Tn2	4.8	RSF1010	Ap	random	88
Tn3	4.6	R1drd19	Ap	random	92
Tn4	20.5	R1drd19	Ap Sm Su	random	92
Tn5	5.3	JR67	Km	random	86
Tn6	4.1	JR72	Km		86
Tn7	5.6	R483	Tp Sm/Sp	specific (few sites)	90
Tn8[b]	–	–	–	--	-
Tn9	2.4	pSM14	Cm		89
Tn10	9.4	R100	Tc	nonrandom	87,93

[a]Abbreviations: Ap, ampicillin; Cm, chloramphenicol; Km, kanamycin; Sm, streptomycin; Sp, spectinomycin; Tc, tetracycline; Tp, trimethoprim; Su, sulfonamide.

[b]Tn8 has not been described.

that can easily be scored. Transposition element Tn1 might be used along with Tn5 as a cointegrate. Then β-lactamase activity expressed by Tn1 could be scored.

Transposition elements have the added advantage of inactivating genes upon insertion. Spontaneous loss of the element restores the gene's functional activity. Thus insertionally inactivated genes may also be inserted into protoplasts and clones can be screened for spontaneous loss of the element and a gain in genetic functional expression.

It is not known if similarities exist between prokaryotic transposon sequences and those in higher plants. If prokaryotic and eukaryotic transposition elements are similar, then there exists the possibility of integrating foreign genes into plant DNA.

Acknowledgments: Portions of this manuscript were written during the author's stay in the laboratory of Paul Lurquin, Centre D'Étude de L'Énergie Nucléaire, Mol, Belgium. The author is indebted for the many stimulating discussions during his visit. Special thanks to Paul Lurquin and Carrie Ireland for reading the manuscript. Work in the author's laboratory cited in this article was supported in part by research grants from the National Cancer Institute, NIH, and North Atlantic Treaty Organization.

REFERENCES

1   Lurquin, P.F. (1977) Prog. Nucl. Acid Res. Mol. Biol. 20,
    161-207.
2   Kleinhofs, A. and Behki, R. (1977) Ann. Rev. Genet. 11, 79-101.
3   Kado, C.I. (1976) Ann. Rev. Phytopathol. 14, 265-308.
4   Ohyama, L. (1978) in Proc. 4th Int. Congr. Plant Tissue Cell
    Culture, Calgary (in press).
5   Lurquin, P.F. and Kado, C.I. (1977) Mol. Gen. Genet. 154,
    113-121.
6   Hotta, Y. and Stern, H. (1971) in Informative Molecules in
    Biological Systems (Ledoux, L.G.H., ed.), pp. 176-186, North
    Holland, Amsterdam.
7   Shepherd, R.J., Bruening, G.E. and Wakeman, R.J. (1970)
    Virology 41, 339-347.
8   Brunt, A.A. (1971) Ann. Appl. Biol. 67, 357-369.
9   Szeto, W.W., Hamer, D.H., Carlson, P.S. and Thomas, C.A. (1977)
    Science 196, 210-212.
10 Shepherd, R.J. (1976) Adv. Virus Res. 20, 305-339.
11 Fujisawa, I. and Matsui, C. (1972) Virus 22, 30-32.
12 Fujisawa, I., Rubio-Huertos, M., Matsui, C. and Yamaguchi, A.
    (1967) Phytopathology 57, 1130-1132.
13 Shepherd, R.J. and Wakeman, R.J. (1971) Phytopathology 61,
    188-193.
14 Pirone, T.P., Pound, G.S. and Shepherd, R.J. (1961) Phyto-
    pathology 51, 541-546.
15 Favali, M.A., Bassi, M. and Conti, G.G. (1973) Virology 53,
    115-119.
16 Meagher, R.B., Shepherd, R.J. and Boyer, H.W. (1977) Virology
    80, 362-375.
17 Fujisawa, I., Rubio-Huertos, M. and Matsui, C. (1972) Phyto-
    pathology 62, 810-811.
18 Fujisawa, I., Rubio-Huertos, M. and Matsui, C. (1974) Phyto-
    pathology 64, 287-290.
19 Goodman, R.M. (1977) Virology 83, 171-179.
20 Harrison, B.D., Barker, H., Bock, K.R., Guthrie, E.J.,
    Meredith, G. and Atkinson, M. (1977) Nature 270, 760-762.
21 Brunt, A.A. and Kitajima, E.W. (1973) Phytopathol. Z. 76,
    265-275.
22 Sarkar, S. (1976) Virology 70, 265-273.
23 Kitajima, E.W., Betti, J.A. and Costa, A.S. (1973) J. Gen.
    Virol. 20, 117-119.
24 Kojima, M., Shikata, E., Sugawara, M. and Marayama, D. (1969)
    Virology 39, 162-174.
25 Kellenberger, E. and Edgar, R.S. (1971) in The Bacteriophage
    Lambda (Hershey, A.D., ed.), pp. 271-295, Cold Spring Harbor
    Laboratory, Cold Spring Harbor, NY.
26 Kellenberger, G., Zichichi, M.L. and Weigle, J.J. (1961) J. Mol.
    Biol. 3, 399-408.

27   Davidson, N. and Szybalski, W. (1971) in The Bacteriophage
      Lambda (Hershey, A.D., ed.), pp. 45-82, Cold Spring Harbor
      Laboratory, Cold Spring Harbor, NY.

28   Thomas, M., Cameron, J.R. and Davis, R.W. (1974) Proc. Nat.
      Acad. Sci. U.S.A. 71, 4579-4583.

29   Salivar, W.O., Tzagoloff, H. and Pratt, D. (1964) Virology
      24, 359-371.

30   Davis, R.W., Thomas, M., Benton, D., Cameron, J., Philippsen,
      P., Struhl, K., St. John, T. and Kramer, R. (1977) in Molecular
      Cloning of Recombinant DNA (Scott, W.A. and Werner, R., eds.),
      pp. 155-159, Academic Press, New York, NY.

31   Emmons, S.W., MacCosham, V. and Baldwin, R.I. (1975) J. Mol.
      Biol. 91, 133-146.

32   Scandella, D. and Arber, W. (1976) Virology 69, 206-215.

33   Jones, K.W. and Murray, K. (1975) J. Mol. Biol. 96, 455-460.

34   Kramer, R.A., Cameron, J.R. and Davis, R.W. (1976) Cell 8,
      227-232.

35   Grunstein, M. and Hogness, D. (1975) Proc. Nat. Acad. Sci.
      U.S.A. 72, 3961-3965.

36   Sanzey, B., Mercereau, O., Ternynck, T. and Kourilsky, P.
      (1976) Proc. Nat. Acad. Sci. U.S.A. 72, 3394-3397.

37   Bahl, C.P., Marians, K.J., Wu, R., Stawinsky, J. and Narang,
      S.A. (1977) Gene 1, 81-92.

38   Cohen, S.N., Cabello, F., Casadaban, M., Chang, A.C.Y. and
      Timmis, K. (1977) in Molecular Cloning of Recombinant DNA
      (Scott, W.A. and Werner, R., eds.), pp. 35-56, Academic Press,
      New York, NY.

39   Scheller, R.H., Dickerson, R.E., Boyer, H.B., Riggs, A.D. and
      Itakura, K. (1977) Science 196, 177-180.

40   Messing, J., Gronenborn, B., Müller-Hill, B. and Hofschneider,
      P.H. (1977) Proc. Nat. Acad. Sci. U.S.A. 74, 3642-3646.

41   Miller, J.H. (1972) Experiments in Molecular Genetics, Cold
      Spring Harbor Laboratory, Cold Spring Harbor, NY.

42   Cohen, S.N., Chang, A.C.Y., Boyer, H.W. and Helling, R. (1973)
      Proc. Nat. Acad. Sci. U.S.A. 70, 3240-3244.

43   Hamer, D.H. and Thomas, C.A. Jr. (1976) Proc. Nat. Acad. Sci.
      U.S.A. 73, 1537-1541.

44   Hershfield, V., Boyer, H.W., Lovett, M., Yanofsky, C. and
      Helinski, D. (1974) Proc. Nat. Acad. Sci. U.S.A. 71, 3455-3461.

45   So, M., Gill, R. and Falkow, S. (1975) Mol. Gen. Genet. 142,
      239-249.

46   Hershfield, V., Boyer, H.W., Chow, L. and Helinski, D.R. (1976)
      J. Bacteriol. 126, 447-453.

47   Covey, C., Richardson, D. and Carbon, J. (1976) Mol. Gen. Genet.
      145, 155-158.

48   Rodriguez, R.L., Bolivar, F., Goodman, H.M., Boyer, H.W. and
      Betlach, M.C. (1976) in Molecular Mechanisms in the Control
      of Gene Expression (Nierlich, D.P., Rutter, W.J. and Fox,
      C.F., eds.), pp. 471-477, Academic Press, New York, NY.

49 Bolivar, F., Rodriguez, R.L., Betlach, M.C. and Boyer, H.W.
    (1977) Gene 2, 75-93.
50 Bolivar, F., Rodriguez, R.L., Greene, P.J., Betlach, M.C.,
    Heyneker, H.L. and Boyer, H.W. (1977) Gene 2, 95-113.
51 Figurski, D., Meyer, R., Miller, D.S. and Helinski, D.R. (1976)
    Gene 1, 107-119.
52 Gryczan, T.J. and Dubnau, D. (1978) Proc. Nat. Acad. Sci. U.S.A.
    75, 1428-1432.
53 Lofdahl, S., Sjostrom, J.E. and Philipson, L. (1978) Gene 3,
    149-159.
54 Lofdahl, S., Sjostrom, J.E. and Philipson, L. (1978) Gene 3,
    161-172.
55 Maxam, A.M. and Gilbert, W. (1977) Proc. Nat. Acad. Sci. U.S.A.
    74, 560-564.
56 Kado, C.I. and Lurquin, P.F. (1978) in Microbiology 1978
    (Schlessinger, D., ed.), pp. 231-234, Amer. Soc. Microbiol.
57 Lurquin, P.F. and Kado, C.I. (1977) Arch. Int. Physiol.
    Biochim. 85, 999-1000.
58 Lurquin, P.F. (1978) EMBO Workshop on Plant Tumor Research,
    p. 32, Noordwijkerhout, The Netherlands.
59 Fernandez, S.M., Lurquin, P.F. and Kado, C.I. (1978) FEBS
    Lett. 87, 277-282.
60 Liu, S.T., Lurquin, P.F. and Kado, C.I. (1978) (unpublished
    data).
61 Maliga, P., Sz.-Breznovits, A. and Marton, L. (1973) Nature
    244, 29-30.
62 Maliga, P., Sz.-Breznovits, A., Marton, L. and Joo, F. (1975)
    Nature 255, 401-402.
63 Bourgin, J.-P. (1978) Mol. Gen. Genet. 161, 225-230.
64 Maliga, P. (1976) in Cell Genetics in Higher Plants (Dudits, D.,
    Farkas, G.L. and Maliga, P., eds.), pp. 59-77, Publishing House
    of the Hungarian Academy of Sciences, Budapest.
65 Widholm, J.M. (1977) in Plant Tissue Culture and Its Bio-
    Technological Applications (Barz, W., Reinhard, E. and Zenk,
    M.H., eds.), pp. 112-122, Springer, Berlin-Heidelberg-New York.
66 Struhl, K., Cameron, J.R. and Davis, R.W. (1976) Proc. Nat.
    Acad. Sci. U.S.A. 73, 1471-1475.
67 Sheikholeslam, S., Okubara, P.A., Lin, B.-C., Dutra, J.C. and
    Kado, C.I. (1978) in Microbiology 1978 (Schlessinger, D., ed.),
    pp. 132-135, Amer. Soc. Microbiol.
68 Merlo, D. and Nester, E.W. (1977) J. Bacteriol. 129, 76-80.
69 Sheikholeslam, S., Lin, B.-C. and Kado, C.I. (1978) Phyto-
    pathology (in press).
70 Chilton, M.-D., Drummond, M.H., Merlo, D.J., Sciaky, D.,
    Montoya, A.L., Gordon, M.P. and Nester, E.W. (1977) Cell 11,
    263-271.
71 Chilton, M.-D., Drummond, M.H., Gordon, M.P., Merlo, D.J.,
    Montoya, A.L., Sciaky, D., Nutter, R. and Nester, E.W. (1978)
    in Microbiology 1978 (Schlessinger, D., ed.), pp. 136-138,
    Amer. Soc. Microbiol.

72 Drummond, M.H., Gordon, M.P., Nester, E.W. and Chilton, M.-D. (1977) Nature 269, 535-536.
73 Ledeboer, A.M., Hille, J., Santbulte, W., Ooms, G., den Dulk-Ras, H. and Schilperoort, R.A. (1978) EMBO Workshop on Plant Tumor Research, p. 26, Noordwijkerhout, The Netherlands.
74 de Picker, A., Hernalsteens, J.P. and Schell, J. (1978) (personal communication).
75 Petit, A., Delhaye, S., Tempe, J. and Morel, G. (1970) Physiol. Veg. 8, 205-213.
76 Klapwijk, P.M., Hooykaas, P.J.J., Kester, H.C.M., Schilperoort, R.A. and Rörsch, A. (1976) J. Gen. Microbiol. 96, 155-163.
77 Montoya, A.L., Chilton, M.-D., Gordon, M.P., Sciaky, D. and Nester, E.W. (1977) J. Bacteriol. 129, 101-107.
78 Goldmann, A. (1977) Plant Sci. Lett. 10, 49-58.
79 Lippincott, B.B., Birnberg, P.R., Rao, S. and Lippincott, J.A. (1978) EMBO Workshop on Plant Tumor Research, p. 27, Noordwijkerhout, The Netherlands.
80 Meins, F. Jr. (1974) in Developmental Aspects of Carcinogenesis and Immunity, 32nd Symp. Soc. Develop. Biol., pp. 23-29, Academic Press, New York, NY.
81 Syono, K. and Furuya, T. (1974) Plant Cell Physiol. 15, 7-17.
82 Bednar, T.W. and Linsmaier-Bednar, E.M. (1971) Proc. Nat. Acad. Sci. U.S.A. 68, 1178-1179.
83 Kado, C.I. (1977) in Beltsville Symp. Agr. Res. I. Virology in Agriculture (Romberger, J.A., gen. ed.), pp. 247-266, Allanheld, Osmun & Co., New Jersey.
84 Cohen, S.N. (1976) Nature, 263, 731-738.
85 Hedges, R.W. and Jacob, A.E. (1974) Mol. Gen. Genet. 132, 31-40.
86 Berg, D.E., Davies, J., Allet, B. and Rochaix, J.-D. (1975) Proc. Nat. Acad. Sci. U.S.A. 72, 3628-3632.
87 Kleckner, N., Chan, R.K., Tye, B.-K and Botstein, D. (1975) J. Mol. Biol. 97, 561-575.
88 Heffron, F., Rubens, C. and Falkow, S. (1975) Proc. Nat. Acad. Sci. U.S.A. 72, 3623-3627.
89 Gottesman, M.M. and Rosner, J.L. (1975) Proc. Nat. Acad. Sci. U.S.A. 72, 5041-5045.
90 Barth, P., Datta, N, Hedges, R.W. and Grintner, N.J. (1976) J. Bacteriol. 125, 800-810.
91 Foster, T.J., Howe, T.G.B. and Richmond, K.M.V. (1975) J. Bacteriol. 124, 1153-1158.
92 Kopecko, D.J. and Cohen, S.N. (1975) Proc. Nat. Acad. Sci. U.S.A. 72, 1373-1377.
93 Jorgensen, R., Berg, D. and Reznikoff, W. (1978) in Microbiology 1978 (Schlessinger, D., ed), pp. 181-183, Amer. Soc. Microbiol.
94 Cameron, J.R., Philippsen, P. and Davis, R.W. (1977) Nucl. Acids Res. 4, 1429-1448.
95 Hinnen, A., Hicks, J.B. and Fink, G.R. (1978) Proc. Nat. Acad. Sci. U.S.A. 75, 1929-1933

96   Carbon, J., Ratzkin, B., Clarke, L. and Richardson, D. (1977)
      in Molecular Cloning of Recombinant DNA (Scott, W.A. and
      Werner, R., eds.), pp. 59-69, Academic Press, New York, NY.
97   Sutcliffe, J.G. (1978) Proc. Nat. Acad. Sci. U.S.A. 75,
      3737-3741.

SOYBEAN UREASE-POTENTIAL GENETIC MANIPULATION OF AGRONOMIC IMPORTANCE

J.C. Polacco, R.B. Sparks Jr.,
Department of Genetics

and

E. A. Havir
Department of Biochemistry

The Connecticut Agricultural Experiment Station
P.O. Box 1106, New Haven, Connecticut 06504

INTRODUCTION

Soybean urease is an attractive system for two reasons: 1) it may offer a means of employing tissue culture for improving soybean nutritional value, and 2) it is a possible marker in asexual gene transfers. Recombinant DNA technology plays a crucial role in both of these goals.

Urease is a seed protein common to much of the Leguminosae and first came to our attention as a possible means of improving the nutritional quality of soybean protein. Soy protein contains only 1.2 to 1.3% methionine (1,2; values expressed as grams methionine in a hydrolysate of 100 grams protein), an essential dietary amino acid for humans and monogastric animals. Supplementation of textured soy protein (1.2% methionine) with 1% DL-methionine brings its nutitional value closer to that of beef (1). Amino acid analyses reported for jack bean urease (3,4) translate to methionine contents of 3.9% and 3.7%, respectively. A first assumption was, if the methionine content of soybean urease were equal to that of jack bean urease, then large increases in seed urease could significantly improve overall seed methionine content and nutritive value. It has been reported that urease is 1% of the protein of some jack bean meals (5), while our best soybean variety contains no more than 0.1% urease. It thus seems feasible to breed an extremely high urease trait into soybeans. We have chosen a cellular plant breeding program which involves a second basic assumption—that the urease produced by cultured soybean cells is identical to that found in soybean seeds. Basically, the cellular approach involves: 1) a characterization of the synthesis and regulation of urease in soybean cell cultures, 2) isolation of urease regulatory mutants, and 3) regeneration of high urease cell lines to intact plants to assess the expression and transmission of the high urease trait in the seed.

Besides the obvious danger in expecting a regulatory change in tissue culture (high urease) to result in a developmental change in the intact plant (more urease deposition in developing seeds) the cellular approach suffers from a more serious flaw--soybean plants cannot be regenerated from established cultures. The difficulty in regenerating many crop plants from culture is probably the biggest obstacle in the use of cell and tissue culture as a tool in crop improvement. It seems attractive then to expand the in vitro approach to intact seeds, embryos, meristems or young regenerable cultures which are genetically engineered to acquire traits induced in culture or isolated from sexually incompatible species. Toward the realization of this goal we are attempting to isolate the urease message, and a DNA fragment coding for it, as well as to develop vectors for its reintroduction into a plant host.

This chapter is a progress report dealing with: 1) the validity of the assumptions that the ureases of jack bean seed, soybean seed and soybean cell culture are identical (especially with regard to methionine contents); 2) some characterizations of urease regulation in cultured soybean cells; 3) initial steps in the isolation of the urease message, and 4) the general "clonability" of soybean DNA in E. coli.

METHODS

Biological Material

Callus and suspension cultures of soybean (Glycine max L. var. Kanrich, Burpee Seed Co., Warminster, PA) were induced from shoot tips of sterile etiolated seedlings as described previously (6). Urease was isolated from the variety Prize (Burpee Seed Co.) because it contained more urease than the other varieties tested and its urease has since been shown to be identical to that of Kanrich.

E. coli K12 strains $C600r_k^-m_k^-$ and HB101 were used in recombinant DNA experiments.

Urease Purification and Assays

Details of urease purification from soybean seeds and tissue culture, and urease assay are given elsewhere (7). Urease activity in polysome fractions was assayed in 50 ml flasks containing a 0.9 x 2.0 cm center well and mouth to accommodate a serum stopper.

The center well contained 50 μl of 9 M monoethanolamine. The flask
outside the well contained a 1 ml reaction mix (pH 7.0) of 100 μg
gelatin, 100 μmole Tris maleate, 1 μmole EDTA, 0.2% sodium azide,
0.1 ml polysome fraction and 500 μmole ^{14}C[urea] (12,000 to 13,000
dpm/μmole). Blanks contained water or gradient buffers (12.5% or
50% sucrose) in place of polysome fractions. After 2 to 5 days
incubation at 30°C, reactions were stopped by the injection of 0.5
ml 2 N H_2SO_4 through the serum stopper. This also drives off
$H^{14}CO_3^-$ which is formed from ^{14}C[urea] hydrolysis and which is
trapped quantitatively by the monoethanolamine in the center well.
After 3 hr the monoethanolamine was pipetted into scintillation
vials containing 4 g 2,5-diphenyloxazole (PPO) and 0.4 g p-bis[2-
(5-phenyloxazole)]-benzene (POPOP) per liter of a toluene (2 parts)/
95% ethanol (1 part) mixture. To correct for quenching, counts
per minute (cpm) were converted to dpm by the external standard
ratio method. A unit of urease will hydrolyze 1 μmole urea in 1
min at 30°C and pH 7.0.

Preparation of Soybean Polysomes

Polysomes were isolated from germinating soybean seeds using
the high salt, alkaline buffers employed by Beachy et al. (8) for
developing soybeans. In addition, heparin (Sigma Chemical Co.,
St. Louis, MO) (500 mg/1) and cycloheximide (1 mM) were added to
the grinding buffer while 100 mg/1 heparin was added to the suspension, pelleting and sucrose gradient buffers. Glassware, mortar,
pestle and centrifuge tubes were autoclaved or dry heated at 165°C
overnight. In addition, before glassware came into contact with
polysome preparations it was coated with dichlorodimethylsilane.
Soybeans were spread 1 cm apart on 10 layers of paper towels soaked
in distilled water. After two days in the dark at 27°C the coats
were removed from seeds with emergent radicles. Fifty grams of
this seed preparation were ground to a homogeneous slurry in a
mortar with 50 ml grinding buffer. The slurry and 10 ml of mortar
washings were centrifuged at 10,000 x g for 15 min. The supernatant was recentrifuged and finally layered over 8 ml pelleting
buffer in 10 to 12 tubes. After spinning 97 min at 40,000 rpm
in a Ti 50 rotor, both layers were aspirated and the pellet and
tube washed twice with 2.5 ml suspension buffer. When pellets were
RNase treated, heparin was omitted from the suspension buffer.

Sucrose Gradient Centrifugation of Polyribosomes

The resuspended pellets (approximately 1.5 ml/25 gm seeds)
were cleared by centrifugation at 1,000 x g for 5 min and 1 ml was
layered on a 15 ml linear 12.5-50% sucrose gradient. After spinning
at 25,000 rpm for 90 minutes in a SW 27.1 rotor, 1 ml fractions were

collected by hand from the bottom of the tube. Urease activity was
assayed as described above and absorbance at 254 nm was determined
on 0.1 ml samples. In some cases, before the clearing spin, half
of the preparation was incubated at 37°C in heparin-free suspension
buffer containing 150 µg/ml RNase A.

Preparation of DNA from Cultured Soybean Cells

Total DNA was purified from cultured soybean cells by preparation and lysis of protoplasts, phenol extraction, incubation with
pancreatic RNase and pronase, and adsorption and elution of DNA on
hydroxylapatite.

Protoplast Formation. Protoplasts were prepared by a modification of the method of Ohyama (9). All operations were performed
aseptically to avoid microbial contamination of the DNA preparations.

Cultured cells were harvested and washed with sterile 0-R3-mn
(R3 medium (6), hormone and sucrose free, containing 0.55 M mannitol)
on sterile Miracloth. Cells were resuspended in 0-R3-mn and incubated for about 5 hr at 30°C with the following filter-sterilized
enzyme solution: hemicellulase (10 mg/ml in 0-R3-mn), cellulysin
(20 mg/ml); macerozyme (10 mg/ml) and pectinase (20 mg/ml). Protoplast formation was followed by light microscopy. To check for
contamination of the protoplast preparation, samples were plated
on LB medium (10) and R3 medium and the plates incubated at 30°C
or 37°C for at least 7 days.

Protoplast lysis and DNA extraction. Protoplasts were harvested at 300 x g and washed with 1 x SSC which contained 0.43 M
mannitol. Protoplasts were then suspended in 1 x SSC containing
3% SDS and 0.01 M EDTA, pH 8.0. Lysis occurred immediately and
the lysate was incubated at room temperature for 1 hr to ensure
complete lysis. An equal volume of phenol (saturated with 0.001 M
Tris-HCl, pH 7.5) was added to the viscous lysate and gently mixed
by inversion for 5 to 20 min. The aqueous layer was removed and
the nucleic acid precipitated by addition of 2 volumes of ethanol.
The nucleic acid was pelleted by centrifugation and resuspended
in sterile TEN (0.05 M Tris-HCl, 0.005 M EDTA, 0.05 M NaCl, pH 8.0)
buffer and dialyzed against TEN until the DNA was in solution.
The dialyzed DNA was incubated with 200 µg/ml pancreatic RNase
(a stock solution of 2 mg/ml in 0.15 M NaCl, pH 5.0, was boiled
for 10 min) for 30 min at 37°C, followed by incubation with
200 µg/ml pronase (a stock solution 2 mg/ml, in 0.01 M Tris-HCl,
pH 7.0 was predigested for 30 min at 37°C) for 1 hr at 37°C. The
DNA was then precipitated with ethanol, centrifuged, resuspended
in TEN and dialyzed against TEN.

Hydroxylapatite. Soybean DNA as purified from protoplasts
was diluted with an equal volume of 0.01 M potassium phosphate
buffer, pH 7.0 (PB). The diluted DNA was mixed gently, batchwise,
with hydroxylapatite (Bio-Gel, DNA grade) that was equilibrated
with PB. The DNA-hydroxylapatite mixture was incubated at room

temperature for 1 hr, gently mixed again, and centrifuged at 4000 rpm in an SS34 rotor for 5 min. The supernatant was removed by aspiration. The hydroxylapatite was washed by mixing gently with 0.15 M PB, allowing it to stand at room temperature for 30 min, centrifuging it, and removing the supernatant. DNA was eluted from the hydroxylapatite with 0.5 M PB and the supernatant was removed with a large bore plastic pipet. The DNA eluate was dialyzed against TEN, precipitated with ethanol, pelleted by centrifugation, washed with ethanol, resuspended in TEN, and dialyzed until the DNA was in solution.

Isolation of E. coli Clones That Contain Soybean DNA

The methods for obtaining clones containing soybean DNA will be presented in detail elsewhere (Sparks, R.B. Jr. and Chisolm, D., in preparation). In summary, the protocol was as follows.

a) Purified soybean DNA was partially digested with restriction endonuclease HindIII, and the restriction fragments were covalently attached with T4 ligase (11) to HindIII restricted plasmid pBR322 DNA (12).

b) E. coli C600$r_k^-m_k^-$ was transformed with the ligated DNA as previously described (13).

c) E. coli cells that contained recombinant plasmids were selected by a modification of the D-cycloserine procedure (14) with 2 cycles of selection instead of one.

d) Plasmid DNA was purified by cesium chloride-ethidium bromide equilibrium centrifugation from cells that survived the D-cycloserine selection and were resistant to ampicillin and sensitive to tetracycline.

e) Purified plasmid DNA was characterized by restriction endonucleases and agarose gel electrophoresis (15), electron microscopy, and hybridization with soybean DNA by the Southern procedure (16).

Miscellaneous Procedures

The preparation of anti-urease antiserum, purified anti-urease antibodies, and purified monospecific anti-urease antibodies has been described (7). Protein was determined by a modification of the Biuret procedure (6) or by the BioRad protein assay kit. Phosphate was determined with a commercial kit (Sigma Chemical Co., St. Louis, MO) or as described previously (7).

RESULTS AND DISCUSSION

Urease From Soybean, Jack Bean Seeds and Soybean Suspension Culture

Urease was purified 500-fold from ground dry soybeans (7). Figures 1 and 2 show the chromatographic behavior of urease in

the last two purification steps. Urease elutes from hydroxylapatite
(Figure 1) as a single protein species. However, the urease peak
eluting from agarose A–15m (Figure 2) actually contains two active
species. These most likely share a homo–n–mer:homo–(n/2)–mer relationship, can be interconverted by changes in buffer ionic strength,
and are electrophoretically separable (7) (Figure 3). It is significant that partially purified urease from soybean suspension
culture contains these same two urease species as shown by assays
of acrylamide gel slices after electrophoresis of a tissue culture

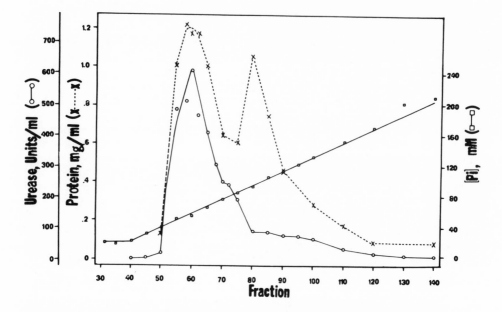

Figure 1. Elution of urease from an hydroxylapatite column.
A 2 x 45 cm column of hydroxylapatite (Hypatite C, Clarkson
Chemical Co., Williamsport, PA) was loaded with a partially
purified urease preparation in pH 7.0 buffer containing 1 mM
EDTA, 1 mM β–mercaptoethanol, and either 10 mM potassium phosphate or 100 mM Tris–maleate. Enzyme was eluted with an 800 ml
linear gradient of 10 to 300 mM potassium phosphate, and 4.3 ml
fractions were collected.

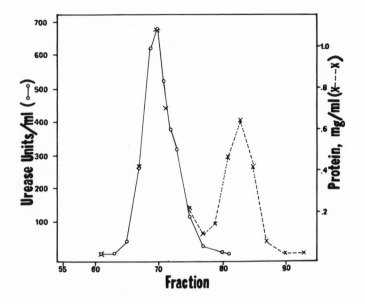

Figure 2. Preparation of pure urease by agarose gel chromatography.
Hydroxylapatite fractions (Figure 1) containing at least 20 urease
units were pooled and concentrated by the addition of $(NH_4)_2SO_4$ to
55% saturation and, after dialysis, applied (total volume, 4 ml)
to a 4 x 45 cm column of agarose A-15m (BioRad, Richmond, CA).
Sample and column were equilibrated with 10 mM KPO_4, 1 mM EDTA,
1 mM β-mercaptoethanol, pH 7.0. Constant flow was maintained by
a peristaltic pump. Fractions of 3.1 ml were collected.

preparation of urease (Figure 3). More convincing evidence for
the identity of soybean seed urease and the urease from tissue
culture is the inhibition of both by purified antibody specific
for soybean seed urease (7). This antibody preparation has been
freed of antibodies that are cross-reactive with jack bean urease
(Figure 4) and is thus a very sensitive probe for similarities
between the ureases of soybean seed and tissue culture.
    Soybean and jack bean urease have identical subunit (93,500
daltons) and multimeric sizes (480,000). They share common
antigenic determinant(s) (Figure 4) and both exist as complexes
of polymeric variants (7). Both are likely to be nickel metallo-
proteins (7,17,18). Most germane to our goals is that they have
identical methionine contents (7).

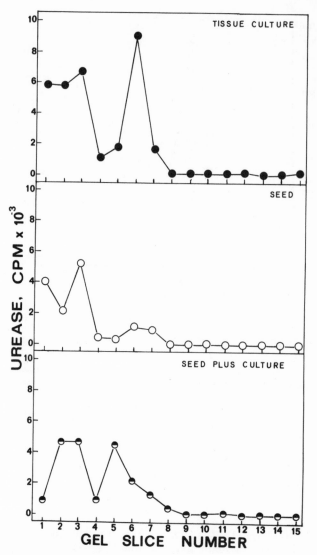

Figure 3. Electrophoretic similarities between the ureases of soybean seed and soybean cell suspension culture. Partially purified
urease from cell suspensions and pure urease from seed were diluted
to 0.15 units/ml and their mobilities examined upon electrophoresis
in polyacrylamide gels (7). Gels were sliced by hand and gel strips
assayed as described for the radioactive assay of urease in polysome
fractions (see Methods). Two species are apparent in both preparations. Peak activities differ by no more than a gel slice, which is
probably within the error of the manual cutting method employed.
There appears to be some cell culture urease activity at the origin.
Cell culture preparations usually leave protein material at the
origin and this can retain added (active) seed urease.

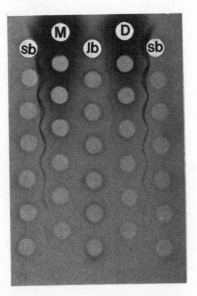

Figure 4. Purification of monospecific anti-soybean urease antibodies. Soybean urease (sb) and jack bean (jb) were added to each
well in the first and fifth column and the third column, respectively
(3 μg/well). Eight micrograms of purified antibodies to soybean
urease were added to the well marked D and successive two-fold
dilutions were added to remaining wells in the column. Cross-
reacting antibodies were separated by passing the preparation three
times over a column of jack bean urease linked to Sepharose 4B.
Eight micrograms of the new preparation were added to well M and
successively diluted two-fold in the remaining wells in the column.
There is no evidence of a precipitin reaction between M (monospecific) antibodies and jack bean urease. However the D (di-
specific) antibodies show a precipitin reaction even when diluted
8-fold.

Thus, the first two basic assumptions for increasing soybean
seed methionine by selecting high urease mutants from cell culture
seem to hold--the urease of soybean seed has the high methionine
content of jack bean urease and is identical to the urease partially
isolated from soybean cell culture.

Regulation of Urease Synthesis in Cultured Soybean Cells

In order to select high urease production in cultured plant
cells it is necessary to understand if and how urease synthesis is
controlled. Our studies to date have demonstrated that urease

levels are both a general function of nitrogen availability and a specific function of urea and nickel availability.

After culturing soybean cells for 1 to 2 days in the absence of any nitrogen source, urease levels drop sharply. The activities in crude extracts of other putative nitrogen assimilatory enzymes are less sensitive to nitrogen deprivation. Glutamic dehydrogenase (6), arginase (17), and glutamine synthetase (unpublished results) are reduced 10, 29 and 65%, respectively, versus reductions of 70 to 100% for urease (6,17) (Figure 5A). Upon addition of either 25 mM urea or 9.4 mM KNO_3 plus 10.3 mM NH_4NO_3, there is an increase in urease production (Figure 5A). The fact that urea addition stimulates greater urease production suggests that urea is a specific inducer of urease, since urea-supported growth is not as great as growth supported by KNO_3 and NH_4NO_3 (Figure 5B). (KNO_3 and NH_4NO_3 are provided in half the amount employed by Murashige and Skoog (19) and are denoted as MS/2.)

When nickel is added to urea-N medium (in the form of a nickel citrate chelate, 10^{-2}mM $NiSO_4$ and 10 mM K citrate, pH 6.0), urea-supported growth is stimulated to the maximal levels observed with the MS/2 nitrogen source. Whether the nitrogen source is arginine (17), MS/2 (KNO_3 plus NH_4NO_3) or urea, nickel supplementation results in 4- to 10-fold higher urease levels. This nickel stimulation of urease activity could be due to de novo protein synthesis or to activation of preexisting apoenzyme. We are currently attempting to distinguish between these two possibilities.

Although the regulation studies are far from complete, the picture that is emerging is that urease synthesis is labile, responsive to a variety of nitrogen signals and to a signal from intracellular nickel. As regulatory effects become elucidated, selection for high urease cell lines can be devised with greater confidence.

Earlier we reported (6) that ammonia, methylammonia and nitrate had repressive effects on urease production and that under proper conditions methylammonia, nitrate and the urease inhibitor, hydroxyurea, could be used as selective agents for high urease cells. The three selections have yielded cell lines all of which were discarded upon further screening either because the variant phenotype was not stable or because cellular urease levels were normal. Unfortunately, the most powerful selection we have devised to date is for urease-negative cells using arsenate as a negative selection agent (in preparation).

Initial Attempts to Isolate the Urease Message

Since seeds are a rich source of urease, we initiated studies on the ontogeny of urease in developing seeds to identify the best stage for urease messenger isolation. However, Sehgal and Naylor (20) reported that jack bean urease levels are very high during germination. If urease is being synthesized in germinating soybeans,

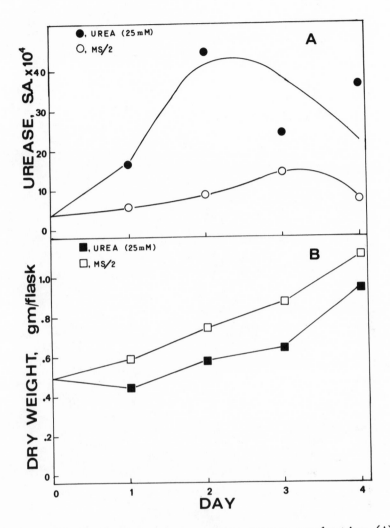

Figure 5. Effect of nitrogen source on urease production (A) and growth (B) in soybean cell suspension cultures. Cells were cultured for 2 days in the absence of a nitrogen source after which replicates (50 ml culture in 250 ml flasks) were made 25 mM in urea or 9.4 mM in KNO_3 and 10.3 mM in NH_4NO_3 (MS/2 (19) nitrogen source) by addition from 100 X stock solutions. Cells were disrupted (7) and urease assayed as described in Methods. A portion of the cells harvested each day were used to determine dry weight yield (65°C, 24 hr).

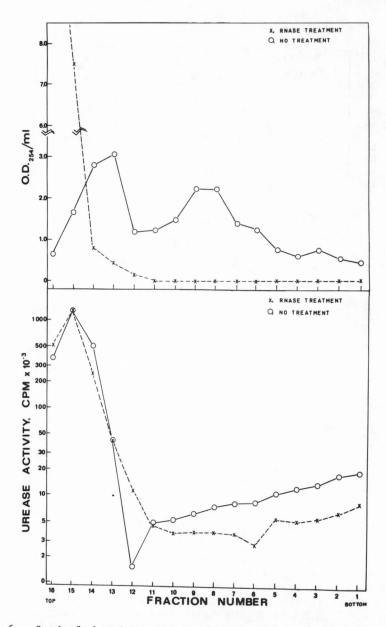

Figure 6. Optical density and urease activity profile of polysomes from germinating soybean seeds. Soybeans were germinated and poly-ribosomes were prepared and sedimented through a sucrose gradient as described in Methods. The polysome portion of the gradient (fraction 1-10) exhibits a low level of ureolytic activity (assayed as described in Methods). At least 50% of this activity does not sediment when polysomes are disrupted by RNase A digestion.

then we could more conveniently use a bag of dry beans rather than
a freezer full of developing beans as starting material.

Soybean seeds were germinated in the dark for 2 days and prepared for polysome isolation as described in Methods (above). Figure 6A is a sucrose density gradient profile of a polyribosome preparation and shows a polysome:monosome ratio >1.5. To find which size
class contain urease message, we employed a sensitive assay for urease catalytic activity in the gradient fractions (see Methods above).
Figure 6B shows that by far the greatest amount of activity is due
to free urease (∿18S) since this sediments above the monosome peak
(∿80S). There is, however, easily detectable ureolytic activity
which sediments with the polysome fractions (1-10; Figure 6B). Half
or more of this activity appears to be associated with polyribosomes
since it will not sediment in RNase-treated preparations (Figure 6B).
It remains to be determined if the activity is due to nascent, ribo-
some-bound urease or to nonspecific aggregation of urease with poly-
ribosomes. Certainly the urease sedimenting in RNase-treated gradients is an aggregate of some kind. We are currently exploring the
use of EGTA (21) as an anti-aggregating agent.

Although not seen in the profile of Figure 6B, ureolytic activity usually appears as a small peak at about fraction 7. We are
currently isolating RNA from different gradient regions to detect
which fractions contain the template for urease synthesis in rabbit
reticulocyte lysates. This system will translate total seed DNA
but it remains to be determined if urease polypeptide will be
synthesized. Optimizing conditions for urease synthesis in vitro
may involve adding nickel since rabbit reticulocyte preparations
are normally treated with EGTA, a potent chelator of nickel. Obviously, our assay of polysomes for nascent urease assumes that it
will have some catalytic activity. It is implicit in this assumption that nickel is bound to the polypeptide before release from
the ribosome. Thus, employing the same basic assumption, we could
likely detect nascent urease activity by soaking soybean seeds with
trace ^{63}Ni (a long-life beta emitter) and directly monitoring
radioactivity in polysome gradients.

Isolation of E. coli Clones Bearing the Soybean Urease Structural Gene

The urease mRNA, when isolated, can be converted to a clonable
cDNA or to radioactive cDNA to use as a probe for shotgunned clones
bearing urease structural gene sequences. Comparison of cDNA and
shotgunned DNA sequences (at least in a preliminary way by heteroduplex analysis) will also indicate if any major processing of the
RNA transcript occurs.

We have tried a less sophisticated approach to select urease-
bearing clones of E. coli: direct expression of urease activity.
E. coli K12 (isolates HB101 and C600) cannot utilize urea. They
are not inhibited by 100 mM urea when glucose is the carbon source
but there is severe growth inhibition when the carbon source is

L-arabinose. Since urea was shown to inhibit the expression of many catabolite repressible operons in E. coli K12 (22), we interpret our finding to mean that urea can enter E. coli cells, where it is a nonusable nitrogen source and nontoxic in the presence of glucose carbon source. To date, we have selected a small number of clones which use urea-nitrogen and have detectable ureolytic activity (unpublished observations); however, this urease-positive phenotype does not persist. We now have urease antibodies and radioiodinated urease to detect low levels of production of urease antigen in putative transformants.

Isolation of DNA from Cultured Soybean Cells

To clone DNA one must first isolate DNA. This modest first step is not always stumble-free when starting with plant material. Problems are often encountered in cell disruption, shearing and contamination by polysaccharides. We have avoided the first two by preparing protoplasts from cultured cells (see Methods above). However, in some cases, soybean DNA prepared from protoplasts was contaminated with a high molecular weight substance, presumably polysaccharide. This material copurifies with DNA up to the hydroxylapatite step. It strongly absorbs ultraviolet light (UV) of 240 nm wavelength and covers the nucleic acid absorbance profile at 254 nm (Figure 7A). The amount of the contaminant varied with each preparation of DNA. In some preparations very little existed, while in others it was overwhelming. The reason for this variation is currently unknown. After the hydroxylapatite step, the DNA that eluted with 0.5 M PB was free of the contaminant (Figure 7B). The hydroxylapatite-purified DNA showed excellent 260:280 and 260:230 ratios (Figure 7B). The 240 nm contaminant was shown to be tightly adsorbed to the hydroxylapatite. After elution of the DNA with 0.5 M PB, the hydroxylapatite was washed with 1 M PB followed by 1 M PB plus 0.05 M EDTA, but the contaminant was not eluted. The hydroxylapatite was then dialyzed against 0.5 M EDTA, pH 8.0, until it dissolved, and dialyzed against TEN for 24 hr. The UV scan of Figure 7C shows that the contaminant was adsorbed to the hydroxylapatite.

The size of the DNA eluted from hydroxylapatite was determined by comparison with known standards electrophoresed on agarose gels. The size of the majority of DNA was over 15×10^6 daltons. However, there was a small amount of DNA material of lower molecular weight which ran the length of an agarose gel. Treatment of purified DNA with restriction endonucleases EcoRl or HindIII did not give a specific pattern of restriction fragments on agarose gels but resulted in a random smear of DNA fragments from untreated size to less than 1×10^5 daltons. Short times of incubation with HindIII gave proportionately higher molecular weight restriction fragments which were used in subsequent cloning experiments.

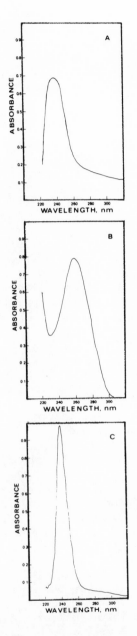

Figure 7. Absorbance profiles of soybean DNA preparation at various stages of purification. Scans were made from 320 nm to 220 nm on a Beckman model 25 scanning spectrophotometer. Panel A, DNA purified up to hydroxylapatite step, 1:4 dilution. Panel B, 0.5 M phosphate buffer eluate from hydroxylapatite. Panel C, Material eluted after dissolving hydroxylapatite in 0.5 M EDTA and dialyzing against TEN, 1:4 dilution.

Partial Characterization of <u>E. coli</u> Clones
Containing Recombinant Plasmids

We have formed <u>in vitro</u> recombinants between soybean DNA and
<u>E. coli</u> plasmid pBR322 DNA. Partial restriction of soybean DNA
with restriction endonuclease <u>HindIII</u>, covalent attachment of the
DNA fragments by T4 ligase to pBR322, and transformation into
<u>E. coli</u> C600$r_k^-m_k^-$ resulted in many clones containing recombinant
plasmids. Cells that survived two cycles of D-cycloserine selection (14) in the presence of tetracycline were tested for resistance
to ampicillin and tetracycline. All surviving cells were ampicillin-
resistant and 50% were tetracycline-resistant. This suggests that
50% of the surviving colonies were recombinants in which the tet^R
gene was inactivated by insertions of DNA in the <u>HindIII</u> restriction site of pBR322.

Preliminary screens for the size of plasmid DNA in 105 amp^R
tet^S clones were performed. Clones were grown in minimal medium
to mid-log phase, chloramphenicol (250 µg/ml) added, and the cells
incubated overnight. The cells were lysed by SDS, the chromosomal
DNA pelleted by centrifugation in an Eppendorf table-top centrifuge,
and the supernatant containing plasmid DNA electrophoresed on
agarose gels. The majority of inserts found in pBR322 (2.6
megadaltons) were less than 2.5 megadaltons. We found that 10 to
15% of the recombinants contained inserts of 2.5 to 6.0 megadaltons.
Several plasmids with inserts of 2.5 megadaltons or greater were
purified by cesium chloride-ethidium bromide centrifugation and
characterized further.

Figure 8 shows an agarose gel pattern of several restriction
endonuclease digestions of plasmid pCS1712. The sum of the sizes
of the restriction fragments gives a plasmid size of 5.2 ± 1.9
megadaltons. This compares well with the size determined by
electron microscopy (5.4 ± 1.6). Thus, the insert has a molecular
weight of about 2.5 x 10^6. DNA-DNA hybridization experiments are
in progress which are designed to show decisively that the inserted fragments are of soybean origin. Very few plasmids with
inserts greater than 7 megadaltons have been isolated, even though
the majority of partially restricted soybean DNA fragments are
larger than this.

It is expected that a DNA fragment of less than 2 megadaltons
would be required to code for a 90,000 molecular weight subunit
of urease. Since soybean DNA restriction fragments of this size
are apparently maintained in <u>E. coli</u>, it is likely that a cloned
cDNA of soybean urease mRNA will also be maintained.

CONCLUSION

Our contribution's place in this volume may seem as dubious to
the reader as it does to the authors. While we certainly have not

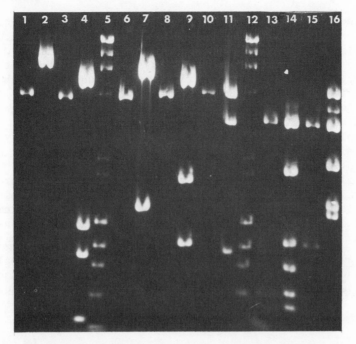

Figure 8. Agarose gel electrophoresis pattern of pCS1712 DNA and
pBR322 DNA restriction fragments. The DNA samples were treated
with restriction endonucleases (BioLabs) SalI (well No. 1,
pBR322; well No. 2 pCS1712), EcoRI (well No. 3, pBR322; well No. 4
pCS1712), PstI (well No. 6, pBR322; well No. 7; pCS1712), BamI
(well No. 8, pBR322; well No. 9, pCS1712), HindIII (well No. 10,
pBR322; well No. 11 pCS1712), HindIII and HincII (well No. 13
pBR322; well No. 14, pCS1712), and HincII (well No. 15, pBR322;
well No. 16, pCS1712) for 3 hr at 37°C. After restriction, the
samples were treated at 65°C for 7 min, mixed with bromphenol blue
and glycerol, and electrophoresed for 8 hr at 20 mAmp on 1%
agarose (Marine Colloids - Sea Kem ME). In wells 5 and 12, a
mixture of HindIII-digested λ DNA and HaeIII-digested φX174 DNA
were electrophoresed as molecular weight markers.

documented a genetic engineering system, we have tried to communi-
cate what we would like to do with a higher plant system via recombinant DNA technology. It also ought to be emphasized that urease
is probably not the only (and, indeed, probably not the best)
methionine-rich protein in the soybean seed. Our approach suggests that biochemically oriented breeding programs can focus on
specific changes to effect agronomic improvement.
      Urease is an ideal natural marker for transformation of
E. coli. E. coli K12 cannot utilize urea nor is any detectable
amount produced by our sensitive assay. However, urea apparently

enters the E. coli cell. Urease is a homo-multimer and does not require a complex prosthetic group so that only the structural gene need be introduced into the E. coli host to enable it to use urea-nitrogen. While the subunit size, 93,500 daltons, appears danger-ously large for faithful transcription and translation, there is strong evidence (3) that the subunit is only 30,000 to 32,000 daltons (with two nickels possibly being stable gluing agents among three subunits).

In spite of the case we have built for the feasibility of transforming E. coli with soybean DNA to utilize urea, we have failed to isolate any transformants. E. coli appears to be capable of maintaining inserts equal to the minimum size of the urease structural gene (0.6 to 1.9 megadaltons, depending on subunit size). It is likely, although certainly not definitely demonstrated, that the soybean recognition sequences for initiating transcription and translation of urease are incompatible with the E. coli milieu.

Acknowledgments: We wish to thank C. Forte, D. Chisholm, M. O'Connell and K. Clarke for excellent technical assistance. M. O'Connell and C. Forte prepared all of the line drawings.

REFERENCES

1   Kies, C. and Fox. H.M. (1971) J. Food Sci. 36, 841-845.
2   Roberts, L.M. (1970) in The Food Legumes, Recommendations for Expansion and Acceleration of Research to Increase Production of Certain of These High Protein Crops, p. 64, The Rockefeller Foundation, New York, NY.
3   Staples, S.J. and Reithel, F.J. (1976) Arch. Biochem. Biophys. 174, 651-657.
4   Milton, J.M. and Taylor, I.E.P. (1969) Biochem. J. 113, 678-680.
5   Varner, J.E. (1959) The Enzymes, 2nd ed., 4, 247-256.
6   Polacco, J.C. (1976) Plant Physiol. 58, 350-357.
7   Polacco, J.C. and Havir, E.A. (1978) J. Biol. Chem. (in press).
8   Beachy, R.N., Thompson, J.F. and Madison, J.T. (1978) Plant Physiol. 61, 139-144.
9   Ohyama, K. (1975) in Plant Tissue Culture Methods (Gamborg, O.L. and Wetter, L.R., eds.), pp. 70-74, National Research Council of Canada, Saskatoon.
10 Miller, J.H. (1972) Experiments in Molecular Genetics, Cold Spring Harbor Laboratory, Cold Spring Harbor, NY.
11 Dugaiczyk, A., Boyer, H.W. and Goodman, H.K. (1975) J. Mol. Biol. 96, 171.
12 Bolivar, F., Rodriguez, R.L., Green, P.J., Betlach, M.C., Heyneker, H.L., Boyer, H.W., Crosa, J.H. and Falkow, S. (1977) Gene 2:95-113.
13 Cohen, S.N., Chang, A.C.Y. and Hsu, L. (1972) Proc. Nat. Acad. Sci. U.S.A. 69, 2110-2114.

14   Rodriguez, R.L., Bolivar, F., Goodman, H., Boyer, H.W. and
      Betlach, M. (1976) in Molecular Mechanisms in Control of
      Gene Expression, ICN-UCLA Symp. Mol. Cell Biol. (Nierlich, D.P.,
      Rutter, W.J. and Fox, C.F., eds.), Vol. 5, pp. 471-477,
      Academic Press, New York, NY.
15   Meyer, R., Figurski, D. and Helinski, D.R., (1977) Mol. Gen.
      Genet. 152, 129-135.
16   Southern, E.M. (1975) J. Mol. Biol. 98, 503-517.
17   Polacco, J.C. (1977) Plant Physiol. 59, 827-830.
18   Dixon, N.E., Gazzola, C., Blakeley, R.L. and Zerner, B. (1975)
      J. Am. Chem. Soc. 97, 4131-4133.
19   Murashige, T. and Skoog, F. (1962) Physiol. Plant. 15, 473-497.
20   Sehgal, P.P. and Naylor, A.W. (1966) Plant Physiol. 41, 567-572.
21   Jackson, A.O. and Larkins, B. (1976) Plant Physiol. 57, 5-10.
22   Sanzey, B. and Ulmann, A. (1976) Biochem. Biophys. Res. Commun.
      71, 1062-1068.

ACKNOWLEDGMENTS

Laine McCarthy was an invaluable help in
editing of this volume. Final processing
of the manuscripts was done by
Laine McCarthy and Maria Beckman.

INDEX

Actin genes, 29
Actinomycin D, effect on cDNA
    synthesis, 21
Adenine phosphoribosyltransferase,
    68,69
Adenovirus DNA, 55,57,110,111
Adenylate kinase, 198
Agrobacteria
    biotype 1 and 2, 157,158
    biotype 3, 158
    mutagenesis, 158
    transformation, 159
    virulence transfer, 161
Agrobacterium radiobacter, 154
Agrobacterium rhizogenes, 151,154,
    157
Agrobacterium rubi, 151,152,154
Agrobacterium tumefaciens, 151,
    152,154,205
Agrocin 84 resistance, 162
    sensitivity, 163
Alkaline phosphatase, 3
α-amanitin, 98,108
Amyloplasts, 182,192
Antirrhinum majus, 183
AP1 exclusion, 162,163
APRT, 68,69
ATP-dependent DNase, 86
AvaI, 98
Avian myeloblastosis virus, 17
Azaserine, 68

Bacteriophage, see also Lambda phage
    f2, 134
    φX174, 147
    M13, 226,228

Bacteriophage (cont'd)
  MS2, 134
  Qβ, 133-143
    A$_2$ protein, 134,139
    coat cistron initiation site,
      141,142
    maturation protein, 134
    replicase, 134,135,138,142
  R17, 134
  RNA phage, 134
BamI, 53-57, 60-63,120,124,184,
    208,212,225,229,230,257
Base analog mutagenesis, 80
Bean, 183
Beta vulgaris, 183
BglI, 85,88,90,184
BglII, 53,230
Bipolaris maydis, 214
Bisulfite mutagenesis, 79,80,87
Blackberry, 152
Blunt end ligation, 7, 38, 86

C3 plants, 191,197
C4 plants, 191,197,198
CaCl$_2$ method, 39
Cane gall, 151,152
Carnation etched ring virus, 226
Carrot, 194
Cassava latent virus, 226
Cauliflower mosaic virus, 205,
    225,226
Chlamydomonas reinhardi, 182,
    187-190
Chloramphenicol, 189
    amplification, 231

Chloroplast, 181–203
  development, 181,182
  DNA, 209
    as site for cytoplasmic male
      sterility, 212
  genome, 182
    number of copies, 196
    reiteration, 184
    renaturation kinetics, 184
    ribosomal protein genes in,
      187
Chorion genes, 29
Chromoplasts, 182,192,199
Clone bank, see Libraries
Cloning, 16
  in Ti-plasmids, 174
  of SV40 mutants, 81,88
Codium fragile, 194
ColEl, 25,118,122,123,125,126,
  227,230,231
Colony hybridization method, 27,
  117,121
Complementary DNA, 15,16
  clones, 15–26
  first strand synthesis, 17
  purity of mRNA template, 18
  second strand synthesis, 21
Complementation plaquing, 81
Conalbumin mRNA, 17
Corn, see Maize and Zea mays
Covalently closed circular DNA, 74
Crop improvement, 196–200
Crown gall, 151,152
Cyclization of linear DNA, 76,77,
  86
Cycloheximide, 189
Cytoplasmic male sterility, 205,
  211–213
  wheat, 212
  sorghum, 212
Cytosine deamination by bisulfite,
  80

Dahlia mosaic virus, 225,226
Daucus carota, 194
Defective SV40, 84
Deletion mutants, generation of
  sequences at deletion joints, 77
DNA isolation from cultured soy-
  bean cells, 254

DNA libraries, 16
DNA polymerase I, 8,17,79,85
  gene amplification, 8
  M. luteus, 79,87
DNA uptake into plants, 224
  by protoplasts, 232
Drosophila 5S DNA, 110,111

E. coli
  complementation by yeast genes,
    117,118
  HB101, 145
  χ1776, 26,27,39
  EcoRI, 7,42,43,53–55,57,60–62,94,
    96,112,123,128,144–146,208,
    212,228–230,257
  site construction, 26
EK2 host-vector, 26
Electrophoresis, preparative, 44,
  45,47
Eliaoplasts, 192
Ethyl methanesulfonate, 158
Euglena gracilis, 183–187
Eukaryote gene expression in
    bacteria, 29
  ovalbumin, 30
  proinsulin, 30
  rat growth hormone, 30
  somatostatin, 30
  yeast genes, 29
Etioplasts, 186
Exonuclease III, 79,87

Ferrodoxin, 199
Fibroin clones, 24
Fibroin mRNA, 18
Folate reductase, 68
Form I DNA, 74
Form II DNA, 74

Gene amplification, 97
Gene dispersal in organelle
    biology, 189,190
Gene enrichment, 37
  cloning, 37
  electrophoresis, preparative,
    44,45,47
  ion-exchange chromatography,
    37
  isopycnic methods, 39

Gene enrichment (cont'd)
  nucleic acid affinity systems,
    40
  R-loop method, 40
  RPC5 chromatography, 40-44,47
Globin clones, 22
Globin gene
  mutagenesis, 144
  purification, 46
Globin mRNA, 16,29
Glusulase, 119
Glutamine synthetase, 199
Golden yellow mosaic virus, 226
Grana, 190
Growth hormone clones, 23
Growth hormone mRNA, 29
Grunstein-Hogness procedure, 117,
  121

HaeII, 98,212
HaeIII, 19,208,257
Hairy root, 151,152
HapI, 112
HAT medium, 52,65
Helianthus annuus, 173
Helminthosporium maydis, 214
Helper-dependent SV40 mutants, 89
Helper virus, 80
Herpes simplex, 52,207
  DNA
    restriction enzyme digestion,
      53,55,61,62
    transformation with tk gene,
      56,57,65
Heterocyst, 199
Heteroduplex mapping, 83
HhaI, 98
HincII, 61-63,257
HindII, 7,76
HindIII, 7,53,54,76,85,86,89,96,
  112,125,184,208,210,229,230,245,
  256,257
his3 transformation, 126,128
his4 transformation, 120,124,127
Histopine, 154,156
Homo-octopine, 166,167
  resistant mutants, 168
HpaI, 53,54,61,98,229,230
HpaII, 79,87
Hybridization-arrested translation,
  28

Hybridization probe, 16
Hydroxycytidine nucleotide, 135,
  136,138,144,145
Hypoxanthine guanosine phosphoribosyl transferase, 52

Imidazole glycol phosphate de-
  hydratase, 123
Immunoglobin clones, 23
Incompatibility group P, 159,162,
  170
  group W, 159
Insulin clones, 23
Insulin mRNA, 29
Intron, 16,114
In vitro packaging, 39
Isoelectric focusing, 67,68
β-isopropyl malate dehydrogenase,
  123

Jack bean urease, 241,249
JR67, 234
JR72, 234

Kalanchoë daigremontiana, 152,
  154,155,173
KAR mutation in yeast, 128
KpnI, 53,62

Lambda phage
  λgt, 226
  λgt-λC, 228
  cloning, 25,38
  exonuclease, 5
Leghemoglobin, 175,199
Lettuce, 183
leu2 transformation, 122,123,
  125-128
Libraries, 16
  identification of sequences
    from, 29
  maize, 193
  yeast, 124
Ligase, 7,86
  E. coli, 7
  T4, 7
Lincomycin, 189
Ltk⁻ clone d, 52
Lysopine, 154,156
  dehydrogenase, 164,165
Lysozyme mRNA, 18

Maize plastid chromosome, 182,183
Maize streak virus, 226
Male-sterile maize, 205
Mapping mutants, 83
Marker rescue, 83,90
Methotrexate, 68,70
2 micron DNA in yeast, 118,127, 128
Mirabilis mosaic virus, 226
Mitochondria
  proteins in yeast, 189
  rRNA, 211
  5S RNA, 211
Mitochondrial DNA, 206
  Aspergillus nidulans, 207
  cucumber, 207
  flax, 207
  heterogeneity, 206,207
  maize, 206-210
  maternal transmission, 216
  Neurospora crassa, 207
  potato, 207
  sorghum, 206,209,210
  soybean, 206,207
  Virginia creeper, 206,207
  wheat, 207
Mouse L cells, 68
mRNA sequence, 16
Mutagenesis
  Agrobacterium, 158
  base analogs, 80
  bisulfite, 79,80
Mutant mapping, 83
Mutations
  at a selected site, 73-92,133-148
  use of ethidium bromide, 74

Nick translation, 8,85
  in mutant construction, 190
Nicotiana, 190
  N. suaveolens, 194
  N. tabacum, 152,194,231
Nitrite reductase, 199
Nitrogen fixation, 198,199
  genes, 174
Nitrogenase, 198,199
Nitrosoguanidine, 158

Nopaline, 154,156
  catabolism, 157,162,163
  dehydrogenase, 164,165
  oxidase, 165,167
  permease, 166,167
  synthesis, 157
Nopalinic acid, 156
Nor-octopine, 166,167
Nuclear restorer genes, 218

Oat, 183
Octopine, 154,156
  catabolism, 157,162,163
    genes, 168
  dehydrogenase, 164
  oxidase, 165,167
  permease, 165,167
  synthesis, 157
    genes, 168
Octopinic acid, 154,156
Oenothera hookeri, 183
Oligo(dG)·oligo(dC)-tailing
  method, 26
Oligo(dT)-cellulose, 19
Opines, 164
Ornaline, 154,156
Ornopine, 156
Ovalbumin clones, 23
Ovalbumin mRNA, 17,29
Oviduct mRNA, 17
Ovomucoid mRNA, 17

pAL208, 171
Papovavirus, 73
pAtAG60, 171
pAtKerr14, 171
pBR313, 124,227,229-231
  in cowpea, 231
  in turnip, 231
pBR322, 25,118,193,230,231,245, 256,257
pCR1, 230,231
pCR10v2.1, 30
pCS1712, 256,257
Pea, 183
pβG1, 25,144-146
PGM16, 230

Phosphoenol pyruvate carboxylase, 191,198
Photorespiration, 192
Photosynthetic membrane proteins, 190
Phyllosticta maydis, 214
Plant breeding, 213
Plant viruses, 225-228
Plasmid
  amplification, 25
  cloning, 25,38
  incompatibility, 170
Plastid genome, 182,183
pMB9, 129,193,227,230
pML21, 193
Polyethylene glycol, 118
Poly(dT)·poly(dA)-tailing method, 17
Polysomes from germinating soybean seeds, 252
pOp230, 30
Potato leafroll virus, 226-228
pOV230, 25,30
Pregrowth hormone
  clones, 23
  mRNA, 29
Preproinsulin
  clones, 23
  mRNA, 29
Primer, 9
Probes for screening, 27,28
pSC101, 230
PSC194, 230
Pseudogene, 103,106
Pseudomonas plasmid, 168
pSM14, 234
PstI, 61,184,230,257
  site construction, 26
pTiB6, 171
pTiC58, 171
PUB110, 230
PVH51, 230
pX1r14, 99,100
pYehis1, 123,124,126-128
pYehis4, 120,124,127
pYeleu10, 122-127
Pyrophosphatase, 198
Pyruvate P$_i$ dikinase, 197

R68.45, 160
R100, 234
R1drd19, 234
R483, 234
R702, 159,160,170
R751.pMG1≠2, 159
rDNA, 93-96
  amplified, 98
  methylation, 98
  transcription in vitro, 101
  transcription in vivo, 98
    capping, 99
    initiation, 99,100
    promoter, 99
Restorer genes, 218
Restriction endonuclease, 5
  class I, 6
  class II, 7
  lack of in plants, 224
Restriction map, 7
  mitochondrial DNA, 210
  SV40, 74,75
  Ti-plasmid, 168
  Zea mays plastid DNA, 184,186
Reverse transcriptase, 9,17
  globin mRNA, 19,20
  inhibition, 18
    gel impurities, 18
    ionic conditions, 19
    length of mRNA, 19
    primer, amount of, 20
    product size, 20
    purity, 18
    reaction conditions, 18
    RNase contamination, 18
    storage buffer, 19
    yield, 20
Reversed genetics, 133-148
Rf3 locus, 218,219
Rhizobium, 198
  R. japonicum, 154,156,159
  R. leguminosarum, 154,157,158, 160
  R. lupini, 154,156
  R. meliloti, 154,157,158,160
  R. phaseoli, 154,157,158,160
  R. trifolii, 154,157,158
    transformation, 159,160,173

Ribosomal protein genes, 187
Ribulose bisphosphate carboxyl-
     ase, 185,190-192
  gene, 185,186,197
  in *Nicotiana*, 190
RK2::Mu, 230
R-loop method, 40
RNA phages, *see* also Bacterio-
     phage, 134
RNA polymerase I, 101,108
RNA polymerase II, 108
RNA polymerase III, 99,108,
     110-113
  yeast, 109
RNase T_1 fingerprinting, 138,139
RP4, 159,160,170,234
RPC5 chromatography, 40-44, 47
R-plasmid, 159,163
R-plasmid::Mu cointegrates, 169
RSF1010, 234
RSF1030, 186,193
RSF2124, 230
Rubus, 152

S1 nuclease, 4,17,25,77,86,147
  use in heteroduplex mapping, 83
Saccharomyces cerevisiae trans-
     formation, *see* also Yeast,
     117-132
*Sal*I, 53,54,124,184,186,208,210,
     212,225,227,229,230,257
Scp1, 127,230,231
Screening of clones, 27
  hybridization analysis, 38
  phage, 38,228
S cytoplasm, 215-219
  maternal transmission, 216
  plasmid-like DNAs, 215,216
  reversion to male-fertile, 217,
     218
Self-priming, 17
Silkmoth chorion clones, 24
Simian virus 40, *see* SV40
*Sma*I, 99,100,184,208,212,229,230
Solution hybridization, 58-60
Somatomammotropin clone, 23
Sorghum vulgare, 209
Southern blot procedure, 29,44,119,
     121,126,245
Southern corn leaf blight, 213,214

Soybean protoplasts, 244
Soybean urease, 241-259
Spacer, 94,101,102,106,112,211
  caused by unequal crossing-
     over, 102
  heterogeneity, 96,97
Spinacia oleracea, 183
S-plasmids, 205
5S DNA, 102-112
  oocyte, 102-104,106,107
  promoter, 106
  pseudogene, 103
  repeat unit, 103
  somatic, 102,106,108
  transcription, 108,109
    by oocyte nuclei, 108
    by oocyte nuclei homogenates,
       109
    initiation, 109,112
    promoters, 112
    termination, 109,112
  X. borealis, 103,104,106
  X. laevis, 103-105,108
Strawberry vein banding virus,
     226
Stroma, 191
Sunflower, 152
Surface exclusion, 170
SV40, 7,73-92
  defective, 84
  deletion mutants, 76
  mutants, 81
    separation from helper, 84
  promoter controlling cloned
     eukaryotic DNA, 147

T4 ligase, 86
Tandem duplications, 94
tDNA$^{met}_1$, 112,113
Teosinte, 209
Terminal transferase, 8
Tetrad analysis, 121
Texas cytoplasm, 213
Thylakoid coupling factor, 190
Thylakoid membrane proteins, 190
Thylakoid protein gene, 186,187
Thymidine kinase, 52
  gene location, 54
  isoelectric focusing, 67
  mutant reversion, 65

Ti-plasmids, 161-163,205,231,232
  cloning in, 174
  coding for lipopolysaccharide
    changes, 164
  host range, 172
  nopaline catabolism, 163
  octopine catabolism, 163
  oncogenicity, 163
  phage exclusion, 163
  restriction map, 168
  transmissibility, 163
  tumor induction, 163
Tobacco, see Nicotiana tabacum
Tomato, 152
Transfer RNA genes, 112,113
  in ·plastids, 187
Transformation
  definition, 51
  growth, 52
  of bacteria, 26
    E. coli, 26,27,39
    CaCl₂ method, 39
  of mammalian cells, 51-72
    methotrexate resistance, 68,70
    single copy gene, 63
    thymidine kinase, 52,56,58,
      64-66
      physical state of gene in, 58
      gene frequencies, 58-60
      specific sequences in trans-
        formants, 62
    with calf thymus DNA, 66
    with chicken DNA, 66
    with Dictyostelium DNA, 66
    with Drosophila DNA, 66
    with hamster DNA, 66,68,69
    with human DNA, 66,68,69
    with indigenous cellular genes,
      65
    with mouse DNA, 66,68
    with salmon sperm DNA, 66,68
Transposable elements, see Trans-
  posons
Transposons, 168,169,174,233,234
Trifolium parviflorum, 173
Trifolium pratense, 173
trpl transformation, 129

Urease, 241-259
  antibodies, 249

Urease (cont'd)
  assay, 242
  jack bean, 249
  in polysomes, 252
  messenger, 250
  nickel as stimulator, 250
  purification, 245-247
  soybean cell culture, 249
  soybean seed, 249
  subunits, 258
  urea as inducer, 250

Vaucheria, 194
Vitellogenin genes, 113,114
Vitellogenin mRNA, 113

XbaI, 230
Xbol, 104,106,107,109
Xenopus
  clones, 22
  X. borealis oocyte 5S DNA, 110,
    111
    rDNA, 96
    somatic 5S DNA, 110,111
  X. laevis
    oocyte 5S DNA, 110,111
    rDNA, 93-96
  X. mulleri, 96
  oocyte large nuclei, 101
XhoI, 208,212
Xlol, 104,105

Yeast
  gene expression in E. coli,
    117,118
  2 micron circles, 118,127,128
  5S DNA, 110
  spheroplasts, 118-120
    regeneration, 119
  transformation, 117-132
    integrative, 118
    nonintegrative, 118
    type I,II and III, 126
Yellow leaf blight, 214

Zea mays, 206
  plastid DNA, 182-184
  chloroplast rRNA genes, 185
Zea mexicana, 209
Zea perennis, 209
Zymolyase, 122